Climate Change and Forest Governance

Deforestation in tropical rainforest countries is one of the largest contributors to human-induced climate change. Deforestation, especially in the tropics, contributes around 20 per cent of annual global greenhouse gas emissions and, in the case of Indonesia, amounts to 85 per cent of its annual emissions from human activities. This book provides a comprehensive assessment of the emerging legal and policy frameworks for managing forests as a key means to address climate change.

The authors uniquely combine an assessment of the international rules for forestry governance with a detailed assessment of the legal and institutional context of Indonesia; one of the most globally important test case jurisdictions for the effective rollout of 'Reduced Emissions from Deforestation and Degradation' (REDD). Using Indonesia as a key case study, the book explores challenges that heavily forested States face in resource management to address climate mitigation imperatives, such as providing safeguards for local communities and indigenous peoples.

This book will be of great relevance to students, scholars and policymakers with an interest in international environmental law, climate change and environment and sustainability studies in general.

Simon Butt is an Associate Professor of Law at the University of Sydney. He is a leading expert in Indonesian law and has published extensively on constitutional, criminal, civil, human rights, commercial and Islamic law in Indonesia.

Rosemary Lyster is Professor of Climate and Environmental Law at the University of Sydney and is a recognised expert in climate and environmental law, with an extensive publication record in these fields.

Tim Stephens is Professor of International Law at the University of Sydney. He is an international lawyer and geographer who has published extensively in the field of international law, with a particular focus on international environmental law.

Routledge Research in International Environmental Law

Climate Change and Forest Governance

Lessons from Indonesia

Simon Butt, Rosemary Lyster & Tim Stephens

Routledge
Taylor & Francis Group

LONDON AND NEW YORK

First published 2015 by Routledge

2 Park Square, Milton Park, Abingdon, Oxon OX14 4RN
711 Third Avenue, New York, NY 10017, USA

Routledge is an imprint of the Taylor & Francis Group, an informa business

First issued in paperback 2016

British Library Cataloguing in Publication Data
A catalogue record for this book is available from the British Library

Library of Congress Cataloging-in-Publication Data
Butt, Simon, author.
Climate change and forest governance : lessons from Indonesia / Simon Butt, Rosemary Lyster & Tim
Stephens.
pages cm. – (Routledge research in international environmental law)
Includes bibliographical references and index.
ISBN 978-1-138-83362-3 (hbk : alk. paper) – ISBN 978-1-315-73535-1 (ebk : alk. paper) 1. Climatic
changes–Law and legislation–Indonesia. 2. Forestry law and legislation–Indonesia. 3. Forest degradation–
Indonesia. 4. Reducing Emissions from Deforestation and Forest Degradation (Program) 5. Greenhouse
gas mitigation–Law and legislation–Indonesia. 6. Carbon offsetting–Law and legislation–Indonesia.
7. Deforestation–Control–Indonesia. 8. Conflict of laws–Liability for environmental damages–Indonesia.
9. Environmental law, International. I. Lyster, Rosemary, author. II. Stephens, Tim (Associate professor),
author. III. Title.
KNW3336.B88 2015
346.59804'675–dc23
2014032908

ISBN: 978-1-138-83362-3 (hbk)
ISBN: 978-1-138-28161-5 (pbk)

Typeset in 11/12 Garamond 3 LT Std by
Servis Filmsetting Ltd, Stockport, Cheshire

Contents

Table of legislation

Foreign materials

Australia

European Union

Indonesian materials: Legislation

Indonesian materials: Regulations

Table of cases

Constitutional Court decisions

List of abbreviations

ACC	Anti-corruption Court (*Pengadilan Tindak Pidana Korupsi*)
ADP	Ad Hoc Working Group on the Durban Platform for Enhanced Action
ASEAN	Association of Southeast Asian Nations
AWG-KP	Ad Hoc Working Group on Further Commitments for Annex I Parties under the Kyoto Protocol
AWG-LCA	Ad Hoc Working Group on Long-Term Cooperative Action under the UNFCCC
BAL	Basic Agrarian Law
BPHN	National Legal Development Institute Team
BUMD	*badan usaha milik daerah* (regional state-owned enterprises)
BUMN	*badan usaha milik negara* (state-owned enterprises)
CCB	climate, community and biodiversity
CDM	Clean Development Mechanism
CERs	certified emission reduction units
CO_{2e}	carbon dioxide equivalent
COONAPIP	National Coordinating Body of Indigenous Peoples of Panama
COP	Conference of the Parties
COP13	Thirteenth Conference of the Parties
COP17	Seventeenth session of the Conference of the Parties
COP18	Eighteenth Conference of the Parties
CSOs	civil society organisations
DRC	Democratic Republic of the Congo
DPR	*Dewan Perwakilan Raykat* (Indonesia's national parliament)
ERPA	Emissions Reduction Purchase Agreement
EU	European Union
FAO	Food and Agriculture Organisation of the United Nations
FCMC	Forest Carbon Management Concession (*Izin Penyelenggaraan Karbon Hutan*)
FCPF	forest carbon partnership facility
FFI	Fauna & Flora International
FIELD	Foundation for International Environmental Law and Development

FLEGT	(EU) Forest Law Enforcement, Governance and Trade
FOI	freedom of information
FPIC	free, prior and informed consent
GHG	greenhouse gas
GOFG-GOLD	Global Observation for Forest Cover and Land Dynamics
HA	*hutan alam* (natural forest)
HD	*hutan desa* (village forest)
HIR	*Herziene Indonesisch Reglement* (Code of Civil Procedure in force in Java and Madura)
HKM	*hutan kemasyarakatan* (community forest)
HR	*hutan rakyat* (community forest)
HTHR	*hutan tanaman hasil reboisasi* (reforested forest)
HTI	*hutan tanaman industri* (plantation forest)
HTR	*hutan tanaman rakyat* (community plantation forest)
IAFCP	Indonesia–Australia Forest Carbon Partnership
ICCPR	International Covenant on Civil and Political Rights
ICW	Indonesian Corruption Watch
IEA	International Energy Agency
IETA	International Emissions Trading Association
ILO	International Labour Organisation
IPCC	Intergovernmental Panel on Climate Change
IPKTM	*izin pemungutan kayu tanah milik* (extract timber from private land)
IUPHHBK	*izin usaha pemanfaatan hasil hutan bukan kayu* (non-timber forest production utilisation licence)
IUPHHK	*pemanfaatan hasil hutan kayu* (forest product exploration)
JPT	*jatah penebangan tahunan* (annual harvesting quotas)
KFCP	Kalimantan Forests and Climate Partnership
KORPRI	*Korps Pegawai Republik Indonesia* (Public Servant Corps)
KPK	*Komisi Pemberantasan Korupsi* (Anti-corruption Commission)
KUHAP	*Kitab Undang-undang Hukum Acara Pidana* (Code of Criminal Procedure)
LULUCF	land use, land use change and forestry
MA	*Mahkamah Agung* (National Supreme Court)
MOHA	Ministry of Home Affairs
MOP	Meeting of the Parties
MRV	monitoring, reporting and verification
N_2O	Nitrous Oxide
NAMAs	nationally appropriate mitigation actions
NF_3	nitrogen trifluoride
NGO	non-government organisation
OAR	options assessment report
OECD	Organisation for Economic Co-operation and Development
PERADIN	Persatuan Advokat Indonesia
PFC	perfluorocarbons

PK	*peninjauan kembali* (Supreme Court review of 'permanently binding' decisions)
POKJA PI	*Kelompok Kerja Pengendalian Perubahan Iklim* (Forestry Ministry's Climate Change Working Group)
QELROs	quantified emission limitation or reduction objectives
RAN-GRK	Presidential Regulation 61 of 2011 on the Action Plan to Reduce National Greenhouse Gas Emissions
RBg	*Reglement Buitengewesten* (Code of Civil Procedure in force in the rest of Indonesia, excluding Java and Madura)
RE	*restorasi ekosistem* (forest ecosystem restoration)
REDD+	Reducing Emissions from deforestation and Forest Degradation, plus forest enhancement
R-PIN	readiness programme idea note
SBI	Subsidiary Body for Implementation
SBSTA	Subsidiary Body for Scientific and Technological Advice
SOEs	state-owned enterprises
TPTI	selective cutting and planting system
UN	United Nations
UNCLOS	United Nations Convention on the Law of the Sea
UNDP	United Nations Development Programme
UNEP	United Nations Environment Programme
UNFCCC	United Nations Framework Convention on Climate Change
UPK4	*Unit Kerja Presiden untuk Pengawasan dan Pengendalian Pembangunan* (Presidential Working Unit for Development Supervision and Management)
VCM	Voluntary Carbon Market
VCS	verified carbon standard
VCU	voluntary carbon unit
VER	verified emission reduction
VPAs	voluntary partnership agreements
VRB	village representative body
WRI	World Research Institute

Glossary

adat	customary law
ajudikasi nonlitigasi	non-litigation adjudication
Aliansi Masyarakat Adat Nusantara (AMAN)	Indigenous People's Alliance of Archipelago
badan publik	public bodies
badan publik negara	State public bodies
badan usaha milik negara (BUMN)	state-owned enterprises
bupati	regents (heads of district-level government)
cassation	appeal (to the Indonesian Supreme Court)
dana reboisasi	reforestation fund (managed by the Forestry Ministry)
dapat	could
demokrasi	democracy
Dewan Perwakilan Raykat (DFR)	Indonesia's national parliament
disahkan	approved (by the State)
ditentukan	stipulated
hak guna bangunan	right to build (on land)
hak guna usaha	right to cultivate (land)
hak hutan	forests that have rights or concessions awarded over them
hak milik	right of ownership/freehold title
hak pakai	right to use (land)
hak pemungutan hasil	forestry product-harvesting concessions
hak pengusahaan huta	forestry exploitation concessions
hak pengusahaan perairan pesisir	coastal water concessions
hak ulayat	communally held land
hakim	judge
Herziene Indonesisch Reglement (HIR)	Code of Civil Procedure in force in Java and Madura
hubungi aku kalau ingin menang	contact me if you want to win (see also *hakim*)

hutan adat	customary forest
hutan alam (HA)	natural forest
hutan desa (HD)	village forest
hutan hak	forestry over which rights have been granted
hutan kemasyarakatan (HKM)	community forest
hutan negara	State forest
hutan rakyat (HR)	community forest
hutan tanaman (HT)	plantation forest
hutan tanaman hasil reboisasi (HTHR)	reforested forest
hutan tanaman industri	industrial plantation forests
hutan tanaman rakyat (HTR)	community plantation forest
izin pemungutan hasil hutan bukan kayu	permission to harvest non-timber products
izin pemungutan hasil hutan kayu	permission to harvest timber products
izin pemungutan kayu tanah milik (IPKTM)	private land
Izin Penyelenggaraan Karbon Hutan	Forest Carbon Management Concession (FCMF)
izin usaha pemanfaatan hasil hutan bukan kayu (IUPHHBK)	non-timber forest production utilisation licence
izin usaha pemanfaatan hasil hutan kayu	permission to exploit timber products
izin usaha pemanfaatan jasa lingkungan	permission to perform environmental services
izin usaha pemanfaatan kawasan	permission to exploit forest areas
izin usaha wisata alam	eco-tourism permit
jatah penebangan tahunan (JPT)	annual harvesting quotas
kabupaten	district
kantor lelang	auction house
kantor pengadilan	house of justice
kasai	cassation (appeal to the Supreme Court in Jakarta)
kawasan hutan	forest zone
keadilan sosial	social justice
kehakiman undang-undang/ peraturan pemerintah	statutes/interim emergency laws
Kelompok Kerja Pengendalian Perubahan Iklim (POKJA PI)	Forestry Ministry's Climate Change Working Group
kemanusiaan yang adil dan beradab	a just and civilised humanity
Kementerian Kehutanan	Ministry of Forestry
Kepala Pusat Hubungan Masyarakat	Director of Public Relations
kepentingan umum	public interest

Ketetapan MPR	Decrees of the People's Consultative Assembly
ketuhanan yang maha esa	belief in unitary deity
Kitab undang-undang Hukum Acara Pidana (KUHAP)	Code of Criminal Procedure
Komisi Ombudsman Nasional	National Ombudsman Commission
Komisi Pemberantasan Korupsi (KPK)	Anti-corruption Commission
Korps Pegawai Republik Indonesia (KORPRI)	Public Servant Corps
kota	city
lex posteriori derogat lex priori	if two laws conflict with each other, the more recently enacted law prevails
lex specialis derogat lex generalis	if two inconsistent laws are applicable, the more specific rule overrules the more general law
mafia peradilan	justice-sector mafia
Mahkamah Agung (MA)	National Supreme Court
masyarakat hukum adat	customary law communities
memperhatikan	to observe (community opinion)
mempunyai kekuatan hukum tetap	to have permanent legal authority (i.e. for a decision to be binding)
menunjuk	designate (land)
negara hukum	State based on law (commonly translated as 'rule of law')
negara kesejahteraan	welfare state
Orde Lama	Soekarno's 'Old Order'
otonomi daerah	regional autonomy
otonomi seluas-luasnya	widest autonomy possible
otonomi yang luas	broad and wide-ranging autonomy
PAN karbon	carbon storage
Pancasila	the 'Five Principles' (Indonesia's national ideology)
pemanfaatan hasil hutan kayu (IUPHHK)	forest product exploration
pembangunan	development
pendapatan negara bukan pajak	non-tax State income
pengadilan agama	religious courts
pengadilan khusus	special courts
pengadilan militer	military courts
pengadilan negeri	district court
pengadilan tata usaha negara	administrative courts
Pengadilan Tindak Pidana Korupsi (ACC)	Anti-corruption Court
pengadilan tinggi	provincial high court

pengadilan umum	general courts
pengukuhan	gazetted (or confirmed)
peninjauan kembali (PK)	Supreme Court review of 'permanently binding' decisions
penyandang hak	right bearers
penyelenggaraan negara	State administration
peraturan bupati	regent regulation
peraturan daerah (also, *perda*)	regional regulations (also, by-law)
peraturan daerah kabupaten/kota	district/city regulations
peraturan daerah propinsi	provincial regulations
peraturan gubernur	governor regulation
peraturan kepala daerah	regulations of heads of regions
peraturan pemerintah	government regulations
peraturan perundang-undangan	written laws
peraturan perundang-undangan yang lebih tinggi	higher law
peraturan presiden	Presidential regulations
peraturan walikota	mayoral regulation
persatuan Indonesia	unity of Indonesia
propinsi	provinces
RAP karbon	carbon sequestration
reformasi	reformation
Reglement Buitengewesten (RBg)	Code of Civil Procedure in force in the rest of Indonesia (excluding Java and Madura)
restorasi ekosistem (RE)	forest ecosystem restoration
Satgas Pemberantasan Mafia Hukum	Legal Mafia Eradication Taskforce
satu atap	'one roof' reforms
sosialisasi	socialisation
stabilitas	political stability
tata urutan peraturan perundang-undanga	hierarchy of laws
terpenting	the most important
tersangka	suspect
Undang-undang Dasar	Constitution
Undang-Undang Nomor 14 Tahun 2008 Tentang Keterbukaan Informasi Publik	Law 14 of 2008 on Disclosure of Public Information
undang-undang pokok kekuasaan	judiciary laws
Unit Kerja Presiden untuk Presiden untuk Pengawasan dan Pengendalian Pembangunan (UPK4)	Presidential Working Unit for Development Supervision and Management
urusan wajib	obligations
urusan yustisi	justice-sector matters
wilayah pertambangan	mining areas

Preface

This book provides a comprehensive assessment of the legal issues surrounding the implementation of REDD+ in Indonesia. REDD+ includes countries' efforts to reduce emissions from deforestation and forest degradation as well as fostering conservation, sustainable forest management, and enhancing forest carbon stocks. The focus on Indonesia provides a lens through which to understand the many difficulties that tropical rainforest developing countries face when engaging with the REDD+ project. In the first place, they must engage in the international REDD+ negotiations under the United Nations Framework Convention on Climate Change (UNFCCC) so as to influence the taking of any decisions that may affect them. Country positions are not necessarily aligned, as some States look forward to participating in a REDD+ international carbon credit market while others eschew the market entirely and prefer a public funding approach. In addition, the REDD+ decisions that have emerged incrementally since 2007 place binding obligations on developing countries to move gradually towards a system where their emissions reduction activities can be internationally verified. As this book shows, this must be achieved while at the same time complying with the REDD+ safeguards. The obligations are, by any measure, significant. The purpose of this book is to shine a light on Indonesia's progress with embracing REDD+ as a national priority for reducing its greenhouse gas emissions. This experience can serve as a useful guide to those who wish to investigate comparatively the implementation of REDD+ in other jurisdictions, albeit that differing economic, social, cultural, legal and political structures will always provide the prevailing context.

Deforestation in tropical rainforest countries is a significant contributor to human-induced climate change. Deforestation, especially in the tropics, contributes around 20 per cent of annual global greenhouse gas (GHG) emissions and, in the case of Indonesia, amounts to 85 per cent of its annual emissions from human activities (Sari et al. 2007: 3), making Indonesia the third highest emitter of GHGs in the world. REDD+ is a key international policy being adopted to reduce these GHGs and was formally adopted under the UNFCCC at the Thirteenth Conference of the Parties (COP 13) in Bali. Since Bali there have been five significant negotiations: COP15 in Copenhagen in

December 2009; COP16 in Cancun in December 2010; COP17 in Durban in December 2011; COP18 in Doha in November 2012; and COP19 in Warsaw in November 2013. Each of these summits has delivered outcomes that have significant ramifications for achieving the goal of keeping global temperature rise below 2°C above pre-industrial temperatures, and for REDD+ in contributing to this objective. This book analyses the outcomes of these COPs with a particular focus on REDD+ and raises key legal questions regarding the implementation of REDD+ in Indonesia.

COPs 16 and 17 were particularly relevant for the purposes of this book as REDD+ countries were requested to take various actions to facilitate the establishment of REDD+ programmes in their jurisdictions. The REDD+ safeguards were also developed and agreed to. These include the requirement that actions must complement or be consistent with the objectives of national forest programmes and relevant international conventions and agreements. Transparent and effective national forest governance structures must be promoted and supported in accordance with national legislation and sovereignty, while respecting the knowledge and rights of indigenous peoples and members of local communities, consistently with the UN Declaration on the Rights of Indigenous Peoples.[1] Actions must also be consistent with the conservation of natural forests and biological diversity and must address the risk of reversals and reduce the displacement of emissions (Decision 1/CP.16, UNFCCC 2011: 2, clause 2). As is demonstrated in this book, while important REDD+ safeguards have been formally incorporated into UNFCCC decisions, it cannot be assumed that their implementation in Indonesia will be either straightforward or necessarily effective.

Indonesia has played a constructive, although not especially prominent, role in international climate negotiations. Indonesia is a party to both the UNFCCC and the Kyoto Protocol and has also lodged nationally appropriate mitigation actions (NAMAs) to reduce emissions under the Cancun Agreements. Indonesia's GHG emissions profile is dominated by emissions from deforestation and this goes a significant distance towards explaining Indonesia's particular interest in REDD+ and the focus on forestry and land use management in its NAMAs. Indonesia's strong interest in the development of the international climate regime is also a function of its high vulnerability to the effects of climate change. As a country located in the tropics, increases in average global temperatures will have significant effects given the current narrow range of climate variability. Also, across Indonesia's vast archipelago many islands and settlements (including Jakarta) are under direct threat of saltwater intrusion and eventual inundation. In international climate negotiations, Indonesia has adopted a stance common to most developing countries, which is to insist on the primary responsibility of developed countries to reduce their GHG emissions. However, Indonesia has tended to be

---

1 *United Nations Declaration on the Rights of Indigenous Peoples*, opened for signature 13 September 2007, UN Doc. A/RES/61/295 (2007) (entered into force 2 October 2007).

more practical than ideological in its attitude to climate negotiations generally and in respect of REDD+ in particular.

Indonesia has had, and continues to have, an influence on the development of the rules and mechanisms under the UNFCCC relating to REDD+ (although arguably it could take a more proactive role in shaping their content). These standards will need to be incorporated into domestic Indonesian law and policy if there is to be effective linkage of an Indonesian REDD+ scheme into effective global mechanisms for emissions abatement from avoided deforestation. There is significant uncertainty about whether, under the Indonesian legal system, international treaties that Indonesia has signed automatically form part of Indonesian law and can thus be directly enforced by the courts. Government officials appear to treat particular treaties as binding and some Indonesian courts appear amenable to using international law as an aid to interpretation or to fill the gaps in Indonesian law. However, for the most part, Indonesian government officials and courts have not acted on international norms until they are transformed into Indonesian law. For REDD+, this means that any international agreements that Indonesia signs, or even ratifies, will be unlikely to be applied or enforced in Indonesia without further domestic regulation – such as by national statute or Presidential regulation.

The dismantling of authoritarianism and its replacement with constitutional democracy has made introducing and maintaining any legal infrastructure to support REDD+ far more difficult in today's Indonesia than ever before. Lawmakers – particularly national legislators – are now constrained in the exercise of their lawmaking powers by the real threat of having their laws reviewed and invalidated by the Constitutional Court. Many lawmaking institutions, both national and regional, appear to have overlapping jurisdiction, claiming authority to regulate the same subject matter. The result is an inordinately complex system, riddled with conflicting laws but with very few, if any, reliable mechanisms to resolve them.

Jurisdictional disputes in forestry matters that have emerged to date and that are likely to emerge in future are both 'horizontal' and 'vertical'. Some have emerged between national government institutions, such as between ministries. However, most of the disputes have arisen, and are likely to arise in future, between tiers of government, such as between the national government and a regional government. There is also significant potential for legal conflict between levels of regional government, such as between provincial and districts/cities or even between different districts/cities. Much is at stake, because with jurisdiction to control and permit exploitation and extraction comes income for these institutions, which may be necessary to supplement inadequate operational budgets and employee salaries.

REDD+ has already emerged as a significant area of jurisdictional conflict and legal uncertainty. The main REDD+ related regulations adopted to date have been issued by the Ministry of Forestry. They do not, however, provide an adequate legal basis for REDD+ activities in Indonesia. Their precise legal form – Ministerial Regulations – lack legal strength vis-à-vis other types of laws in the Indonesian legal system. In particular, it is unclear whether they can override laws issued by regional parliaments. The substance of these laws is also problematic. The language used is general, the coverage minimal, it is unclear how these laws 'fit' together with other laws, and some provisions appear to seek to regulate issues that fall outside the portfolio of the Forestry Minister.

Problematically, the mechanisms for determining which institution has ultimate authority over a particular matter – and hence whose law will prevail over others – are various but, on the whole, largely inadequate. Some types of conflict can be resolved by the Supreme Court, but there are significant shortcomings in its constitutionally delineated jurisdiction, and in the way it has handled these types of judicial review disputes. Other conflicts can be resolved using bureaucratic processes but these have, in practice, been used only to review and invalidate regional laws seeking to impose taxes or user charges. For other types of legal conflict, rules exist to help determine which law should prevail over the other, but no formal mechanisms or institutions – whether bureaucratic or judicial – appear to be available to resolve them. There are simply no clear rules to resolve some types of conflict.

The one relatively functional mechanism is constitutional review provided by the Constitutional Court. Since 2007 the Constitutional Court has issued decisions that could have ramifications for a statutory REDD+ scheme, depending, of course, on the precise nature and scope of that scheme. The most important of these is the *Traditional Forest Community* case (2012). In that case, the Court recognised the customary law rights (*hak adat*) of traditional communities living in forest areas. The Court held that the State, when exercising its control under Article 33(3) of the Constitution, must observe those traditional rights. According to the Court, the State has 'full authority to regulate and decide upon the availability, allocation, exploitation, administration of forests and legal relationships arising therein'. However, in respect of customary law forests, the authority of the State was limited by the customary law of the forest community.

This decision may have the effect of limiting the involvement of the national and local governments, and perhaps even constrain the application of national and local laws, in areas where customary rights exist. Any national REDD+ statute that purports to regulate the rights of traditional communities, therefore, will be susceptible to constitutional challenge and perhaps invalidity. Such traditional communities would have little difficulty in attracting assistance in bringing an application from well-known and very capable non-government organisations. These include the Indigenous People's Alliance of Archipelago (*Aliansi Masyarakat Adat Nusantara*, or AMAN), which has successfully appeared alongside traditional communities in the Constitutional Court in past cases, including in the *Traditional Forest Community* case (2012).

Although, for the most part, principles of public participation and free, prior and informed consent (FPIC) remain poorly respected in Indonesia, one result of this Constitutional Court decision might be to require developers and investors to engage direct.y with local communities to enter into genuine agreements with them about REDD+. Other recent reforms appear likely to assist Indonesia to meet various REDD+ safeguards. For example, Indonesia has made significant progress in establishing an effective freedom of information regime – a key REDD+ safeguard. In 2008 the national parliament enacted Indonesia's first 'freedom of information' statute – Law 14 of 2008 on Disclosure of Public Information (the FOI law).

In February 2011 the Forestry Minister issued Forestry Minister Regulation P.7/Menhut-II/2011 on Public Information Services in the Forestry Ministry to implement the FOI Law in the Forestry Ministry. The Regulation specifies types of public information that must be available at all times, including forest plans and policies, the forest gazettal process, forest conservation areas, forest exploitation and use, the socioeconomic conditions of communities who live around forests, the names of those holding various rights over forest areas and the territorial scope of those rights. Information about eco-tourism permits must also be made available, as must provincial forest clearing data, rehabilitation areas and data on production and distribution of forest timber and non-timber products. The Regulation also establishes some limited exemptions. Although this FOI regime is still relatively new, the Information Commission, which under the Law hears disputes between information seekers and public bodies, has been deciding most cases in favour of information seekers. Few forestry-related cases had emerged at the time of writing.

Another relatively successful reform has been the establishment of a relatively robust anti-corruption framework under the Anti-corruption Law of 1999 (amended in 2001). This is significant because corruption in the forestry sector is notorious in Indonesia and threatens to derail REDD+ projects. This statute provides significant penalties for a variety of defined offences. The Law would clearly apply to anyone giving or receiving a bribe in return for awarding a concession contrary to law, for example.

Most corruption cases are handled by the national police force and then passed on to public prosecutors. However, the Anti-corruption Commission (KPK or *Komisi Pemberantasan Korupsi*), an institution independent of government established in 2003, can investigate and prosecute most corruption cases itself and can take over corruption investigations and prosecutions from police and prosecutors. The KPK has recently indicated that it will soon begin focusing on natural resources cases. If corruption threatened a REDD+ scheme, then legal action could be taken against perpetrators and the chances of success would not be insignificant.

Finally, the legal system appears to allow traditional communities to sue for loss resulting from REDD+ projects, including by launching a class action. Article 1365 of the Indonesian Civil Code provides a broad legal basis for suing to recover loss for an act that contravenes another person's rights or even that is immoral or inappropriate in the circumstances. This ground would likely be available for traditional communities to seek relief if their traditional rights are breached by a REDD+ project (such as if they were excluded from their traditional lands without adequate consultation or recompense). Traditional communities might also be able to seek relief through Indonesia's administrative courts and via the Ombudsman.

The authors wish to acknowledge the funding provided by the Australian Research Council through a Discovery Project Grant (DP1095541), which made possible the original research contained in this book. They are greatly indebted also to their many colleagues in Indonesia who were instrumental in providing access to information about the current institutional, administrative and legal arrangements for implementing REDD+. Many government agencies and non-governmental organisations in Indonesia have given generously of their time to scope the parameters of, and to facilitate, this research. The authors are also grateful to Dr Mark Sapwell at Routledge for seeing the publication of this work to its conclusion.

Simon Butt
Rosemary Lyster
Tim Stephens
Sydney, August 2014

1 Introduction

While in recent times much attention has been placed on the carbon seques-tration benefits of tropical forests, it is also important to remember that they have the potential to deliver far more than carbon sequestration services. Forests also provide valuable local, regional and global ecosystem services ranging from water quality, flood control, soil stability and biodiversity (see Karousakis 2006; Lyster 2012; Heal et al. 2001: 336; Karousakis and Corfee-Morlot 2007) and also support traditional cultures and provide livelihoods to over 1 billion of the world's poorest people. However, it is the linkage of tropical forests with carbon emissions that has transformed tropical forest loss into a subject of international law.

The specific vehicle for this transformation is Reducing Emissions from Deforestation and Forest Degradation, plus forest enhancement (REDD+), a mechanism formally adopted under the United Nations Framework Convention on Climate Change (UNFCCC) at the Thirteenth Conference of the Parties (COP 13) in Bali. The core logic of REDD+ seems quite simple. Developed countries, as the parties primarily responsible for global warming to date, will pay developing countries to keep their forests standing and hence reduce emissions from forest carbon loss. This simplicity, however, masks myriad complexities of international and domestic law and tropical forest governance – all of which underpin the ability of different countries and stakeholders to effectively and equitably implement, and benefit from, REDD+. Nevertheless, if properly implemented, REDD+ projects do have the potential to contribute to the protection of forest livelihoods among forest dependent populations (Peskett et al. 2006: 1).

As this book demonstrates, there are a number of pressing legal issues associated with REDD+ in Indonesia that need greater clarification and resolution before it can be said that REDD+ schemes are, or will be, imple-mented in an equitable manner, or even consistently with the many safe-guards that are now required under the UNFCCC. The focus of this book is on a number of legal institutions that are necessary for ensuring the legitimacy of REDD+, including how the rights of all stakeholders, but particularly indigenous peoples (whether or not they are formally recog-nised by governments) and local communities, will be protected. These rights include the rights to: participate in decision making around REDD+ schemes; have access to information about the establishment, or proposed establishment, of REDD+ projects in Indonesia; and have access to legal remedies to protect their rights. Consequently, having established the international framework for REDD+, this book addresses key legal and governance issues, including:

- the implementation of international law in Indonesian law
- Indonesia's role in the development of international climate law and policy
- the institutional environment for REDD+ in Indonesia, including anti-corruption reform
- the national regulatory framework for REDD+ in Indonesia
- jurisdictional conflicts among Indonesian institutions and their impact on REDD+
- judicial and administrative relief, transparency and REDD+, and
- the Indonesian Constitutional Court and REDD+.

Key concerns about the implementation of REDD+

Before moving to discuss the legal issues surrounding REDD+ in Indonesia, it is necessary to situate Indonesia's REDD+ framework within the broader international context and to identify some of the key concerns about REDD+ projects. The REDD+ safeguards now negotiated under the auspices of the UNFCCC attempt to address many of these. These include: governance; establishing baselines and national reference levels; monitoring, reporting and verification (MRV); rights of tenure; the rights of indigenous peoples and other forest-dependent local communities; and the conservation of biodiversity. It is the authors' view that while these issues are now all included in the UNFCCC REDD+ safeguards, the assumptions made about the implementation of these safeguards are glib and rather naive. It is because of these concerns that the authors focus specifically on the prospects for addressing these concerns in Indonesia – in reality – rather than based on a set of assumptions that they can, and will, be addressed.

The crucial issue of governance

Good governance is essential to any tropical rainforest country government's capacity to effectively formulate and implement REDD+ policies and legislation. Yet forest governance has long been a concern of the international community. The European Union (EU) developed its Forest Law Enforcement, Governance and Trade (FLEGT) Action Plan in 2003, which sets out a range of measures to: tackle illegal logging by improving forest governance; strengthen people's tenure rights; develop a licensing scheme that assures that timber has been legally produced; and establish a system to independently monitor implementation. In spite of this initiative on the part of the EU, it seems that forestry governance remains a considerable concern for REDD+ as indicated by the World Resources Institute (WRI) in February 2009. The WRI released a review of 25 of the Readiness Programme Idea Notes (R-PINs) submitted to the World Bank's Forest Carbon Partnership Facility, discussed later (see David et al. 2009). The WRI analysed the R-PINs with reference to 17 good governance criteria which it believes are vital for any country that wants to participate in REDD mechanisms. These were organised within six basic processes:

1   law and policy development
2   land tenure administration and enforcement
3   forest management
4   forest monitoring
5   law enforcement
6   forest revenue distribution and benefit sharing (David et al. 2009: 2).

The WRI found that although governance issues are generally well recognised in R-PINs, none of the countries' submissions on this issue can be considered comprehensive. This is partly because the R-PINs are meant to be preliminary, but the WRI found that several critical issues were conspicuously missing in R-PINs. Consequently, the WRI notes that:

•   law enforcement challenges require greater attention, particularly with regard to illegal logging and other forest crimes
•   unclear tenure is a major challenge in most countries and responding to this challenge will require much more effort
•   measures to increase policy coherence between sectors, particularly with regards to land use planning, need more attention
•   the adequacy of existing revenue distribution and benefit-sharing mechanisms should inform the development of a payment system under REDD
•   transparency and accountability in forest monitoring systems for REDD need to be emphasised (David et al. 2009: 2–3).[1]

Establishing baselines and national reference levels

There are a number of reasons why REDD+ was excluded from provisions of the Kyoto Protocol, including from the Clean Development Mechanism (CDM). These include concerns about: the risk of leakage,[2] non-permanence,[3]

---

1   For a comprehensive discussion of governance in the context of REDD+, see McDermott (2013).
2   'Leakage' refers to greenhouse gas emissions that occur outside the project boundary, but which are nevertheless attributable to its activities.
3   'Permanence' refers to the possibility that carbon is released into the atmosphere as a result of fire, illegal logging or a change in government.

establishing baselines,[4] additionality,[5] and difficulties associated with monitoring and measurement. There was also a concern that the potential of REDD+ to deliver significant international offset carbon credits might divert attention away from the need to reduce emissions from fossil fuels. From a technical perspective, the implementation of REDD+ policies requires accurate estimates of emissions reductions at the national scale. These must be additional in the sense that they would not have occurred but for the REDD+ policies, are permanent and do not result in 'leakage'. Reliable data are essential for investor confidence, especially where REDD+ is integrated into international carbon markets. Estimating emissions reductions has three essential elements. These are:

- establishing the baseline (or 'business-as-usual' scenario)
- establishing the crediting line (or reference scenario that enumerates the quantity of emissions reduction)
- monitoring actual rates of deforestation and degradation associated with emissions over time (Bond et al. 2009: 26).

The capacity of developing countries to adequately monitor rates of deforestation is elaborated on later. However, for lawyers, the terms 'baseline' and 'reference scenarios' are highly significant as they are the standards against which a country's performance on REDD+ will be measured, against which compensation will be paid, and against which carbon credits will be issued for sale on the international market.[6] All of this assumes that certainty prevails especially since, as discussed above, this is what the carbon trading market requires. Yet setting the baseline and reference scenarios for individual tropical rainforest countries is problematic.

Establishing a baseline is essentially a hypothetical assessment since it represents a counterfactual business-as-usual scenario where the question is: What would have happened had the REDD+ project not been implemented? Credits can only be issued by comparing the difference between emissions in the baseline scenario and emissions abatement resulting from the project. Since the baseline scenario will never occur, it is impossible to predict with certainty what the scenario results would have been. For this reason, baselines should be established conservatively so as to not overestimate the abatement outcomes of a REDD+ project (Kollmuss et al. 2008: 18).

Meanwhile, Piraud has contended that 'reference levels' are also inherently problematic. There are two principal ways of determining reference levels: those designed on an historical basis by considering past trends, which may

---

4 Credits can only be generated for emissions below the 'baseline', i.e. GHG emissions reduction that would not have occurred in the absence of a CDM project.
5 It must be demonstrated that the carbon sequestration would not have occurred without the incentives provided by the project.
6 For a comprehensive discussion, see Avitabile (2013).

or may not be adjusted according to national contexts; and those based on modelled predictions, which aim to take into account a certain number of variables that are considered determinants of the deforestation rate. The difficulty with reference levels designed on a historical basis is that this method does not take into account 'forest transition' phenomena, whereby the level of economic development and resource scarcity might modify rates of deforestation from one period to another. In addition, countries in which large-scale deforestation is already occurring are given a 'premium' in that their reference levels will be so high that such countries will be able to generate a disproportionately high number of credits, compared with countries which have low historical rates of deforestation (Kollmuss et al. 2008: 5). Contrariwise, many countries are reluctant to use 'predictive' scenarios given that the likely rate of deforestation will not only be influenced by relatively predictable factors, such as demography, road building, and annual economic growth rate, but also uncertain phenomena (such as the price of various agricultural commodities on speculative and highly volatile markets) (Kollmuss et al. 2008: 6).

Clearly, like the issues of 'additionality', 'leakage' and 'permanence', the difficulties relating to baselines and reference scenarios will continue to plague the question of whether emissions reductions from REDD+ projects have been accurately measured. Questions will also continue to be asked about whether liable entities under any emissions trading schemes should be allowed to rely on REDD+ offset credits for compliance purposes rather than surrendering a carbon allowance.

Monitoring, reporting and verification: its link to compliance and enforcement

The enforcement of any international agreement and domestic legislation, which might be enacted by REDD+ countries, will rely on the ability of developing countries to monitor rates of deforestation for compliance with, and enforcement of, the law. Carbon credits, whether they are generated by a national government or on a project basis, will depend for their legitimacy and marketability on the proper monitoring of rates of deforestation against the baselines and reference scenarios. Without this, the credits are unlikely to satisfy any rules and modalities that might be developed by the Conference of the Parties (COP).

From a technological perspective, there are many barriers to accurate monitoring of forest emissions. The most commonly used satellite dataset for forest monitoring at the national level is archived Landsat data. Other satellite sensor and optical remote sensing data can be used, but access to these data depends on cost and availability. Other challenges associated with remote sensing technologies include persistent cloud cover, seasonality and topography – which mean that the current Landsat 5 receiving stations fail to cover REDD+ countries in central and west Africa and central America. Also, although some remote sensing datasets are available free of charge, access to the internet is needed to access archived satellite data. Most of Africa experiences low available bandwidth, which means that alternative means of data delivery need to be arranged, such as by mailing hard disks or DVDs. Finally, remote sensing data must be pre-processed for interpretation, which needs geometric and radiometric corrections that REDD+ countries may not have the capacity to undertake (Kollmuss et al. 2008: 22–4).

'Land tenure' or 'resource tenure'?

The UNFCCC REDD+ documentation is replete with references to 'land tenure' and the 'right and interests' of indigenous people and local communities. The reason for this is that whether a public funding approach or a carbon credits approach, discussed later, is adopted for REDD+ it is necessary to decide who 'owns' the carbon within existing tenure systems. Yet UNFCCC documentation glosses over what this really means. It is as if these notions are so self-evident that their definition is not even required in the various sections that carefully provide definitions of a wide range of terms. Yet there are many reasons why the questions of 'tenure' and who 'owns' the carbon are crucial.

First, there are concerns that indigenous people's and local communities' livelihoods, access to resources and other rights, such as cultural rights, will be disrupted where deforestation is substantially reduced or halted. Indigenous people face the risk that governments, companies and conservation non-government organisations (NGOs) will 'zone' forests, thereby creating protected areas, biological corridors, forest reserves and sustainable forest management zones in order to receive REDD+ payments, while excluding or disadvantaging indigenous and traditional communities (Griffiths 2007: 10).[7] Second, there are concerns that insecure tenure may itself promote deforestation, as resource users clear forests in order to show occupation where land claims are contested. Finally, tenure is the basis on which all REDD+ benefit-sharing arrangements will be determined, since whoever has tenure over the forest 'owns' the carbon rights that emanate from that forest.[8]

The point is that, as in many jurisdictions, the international law outcomes under the UNFCCC, and indeed the many domestic arrangements for REDD+, are responding to a voluntary REDD+ carbon market, and pilot or demonstration activities, which jumped ahead of the regulatory responses that are now slowly emerging. It is to these that we first turn.

Public or private funding for REDD+?

It is difficult to estimate the specific funding needs for REDD+, however, in 2009 the Meridian Institute issued an options assessment report (OAR) for

---

7 Griffiths (2007: 11) also provides evidence of the impact of carbon forestry on indigenous people and peasant communities in the Ecuadorian Andes.

8 For a comprehensive discussion of this, see Fisher and Lyster (2013).

the Norwegian government (Meridian Institute 2009). The OAR estimated that, to achieve 50 per cent global reductions in forest emissions, funding in the order of US$15–35 billion per year will be needed (Meridian Institute 2009: 43). However, there has been ongoing philosophical debate about the best financing approach to REDD+. It has been suggested, for example, that public funding schemes are not an adequate response to deforestation and degradation and that, to remedy the missing or incomplete market for forest ecosystem services, a market-based instrument to capture the carbon, and other, values of forests should be developed (Karousakis and Corfee-Morlot 2007). Market advocates maintain that public funding schemes will not be sufficient to generate the required volume of funds to provide attractive and sustained economic incentives for REDD+ (Griffiths 2007). While this issue is unlikely to be resolved one way or another, both funding approaches are being adopted, including in Indonesia.

The public funding approach

A number of multilateral and bilateral approaches to public funding for REDD+ have emerged since 2007.

The Forest Carbon Partnership Facility

An example of a multilateral public funding scheme is the Forest Carbon Partnership Facility (FCPF),[9] launched at COP13 by the World Bank in response to a request by developing and industrialised countries to explore a framework for piloting REDD+ activities. Two separate funds have been established: the Readiness Fund and the Carbon Finance Fund, both of which operate under the charter establishing the FCPF.[10] Under the Readiness Mechanism, participating developing countries prepare for REDD+ by developing and adopting national REDD+ strategies; developing reference emission levels; designing MRV systems; and setting up REDD+ national management arrangements, including the proper REDD+ safeguards. The Readiness Fund currently comprises about US$258 million committed or pledged by 15 public donors, which have each provided at least US$5 million.[11] Currently, 36 countries from Asia, Latin and Central America and Africa are participating in the Readiness Fund, including Indonesia, based on the R-PINs. The R-PINs have to be reviewed by the Participants Committee and independently reviewed by a Technical Advisory Panel. Once selected, REDD+ Country Participants receive grant support to develop a Readiness Plan, which contains a detailed assessment of the drivers of defor-estation and degradation; terms of reference for defining their emissions

9 Available at http://carbonfinance.org/docs/FCPF_Booklet_English_Revised.pdf.
10 Available at http://www.forestcarbonpartnership.org/charter-and-rules-procedure.
11 Available at www.forestcarbonpartnership.org/fcpf.

reference level, based on past emission rates and future emissions estimates; establishing an MRV system for REDD+; and adopting or complementing their national REDD+ strategy. A consultation plan is also part of the Readiness Plan. The two other stages are the Midterm Progress Review and the R-Package Assessment.[12] Progress with Readiness procedures is mapped by the FCPF.[13]

Under the Carbon Finance Mechanism, five countries will be selected to participate in pilot incentive programmes for REDD+ based on a system of compensated reductions. Following participation in the Readiness Mechanisms, the selected countries: (a) must have demonstrated a commitment to REDD+ and have adequate monitoring capacity; (b) must have established a credible reference scenario and options for reducing emissions; and (c) will receive payments for reducing emissions below the reference scenario. Payments will only be made to countries that achieve measurable and verifiable emission reductions.[14] As of June 2013 the Carbon Fund has 11 government, private sector and NGO contributors, with funds totalling US$390 million. The Fund is expected to operate until 2020.[15]

UN-REDD Programme

The UN-REDD Programme was launched in 2008 and brings together the technical expertise of the Food and Agriculture Organization of the United Nations (FAO), the United Nations Development Programme (UNDP) and the United Nations Environment Programme (UNEP). Its aim is to support the implementation of national REDD+ programmes while promoting the informed involvement of all stakeholders, including indigenous peoples and other forest-dependent communities. UN-REDD is supporting 48 partner countries in Africa, the Asia-Pacific and Latin America in their REDD+ readiness efforts. It is providing: (a) direct support to the design and implementation of UN-REDD National Programmes; and (b) complementary support to national REDD+ activities through developing common approaches, analyses, methodologies, tools, data and best practices. For example, UN-REDD has released its *Guidelines on Free, Prior and Informed Consent*[16] and *Guidelines on Stakeholder Engagement in REDD+ Readiness with a Focus on Indigenous Peoples and other Forest-Dependent Communities.*[17] By June 2013 total funding for these two streams of support amounted to US$172.4

---

12 Details about how the Readiness Fund works can be found in FCPF (2013).

13 See the *FCPF Dashboard*: www.forestcarbonpartnership.org/sites/fcp/files/2013/june2013/FCPF%20Readiness%20Progress_June%2015_PC15.pdf.

14 *Ibid.*

15 Available at www.forestcarbonpartnership.org/sites/fcp/files/2013/june2013/Carbon%20Fund-web_1.pdf.

16 Available at www.un-redd.org/Launch_of_FPIC_Guidlines/tabid/105976/Default.aspx.

17 Available at www.un-redd.org/Stakeholder_Engagement/Guidelines_On_Stakeholder_ Engagement/tabid/55619/Default.aspx.

million, primarily from Denmark, Japan, Norway and Spain.[18] UN-REDD is not without its detractors. For example, in February 2013, the National Coordinating Body of Indigenous Peoples of Panama (COONAPIP) resolved to withdraw from the UN-REDD Programme/Panama's National Environment Agency (ANAM)-Panama, because of failures to take into account the minimum human rights standards of the indigenous people of Panama, and because agreements reached for implementation of the programme had been breached.[19]

Indonesia is a partner country with a UN-REDD National Programme. The operational phase 1 of Indonesia UN-REDD Programme, which commenced in October 2009, was closed in October 2012. During that time the Programme worked with relevant government agencies, NGOs/civil society organisations (CSOs), academics and the private sector at the national level; and in its pilot province, Central Sulawesi, in cooperation with a multi-stakeholder REDD+ Working Group. Key outcomes include the development of a methodology for a reference emission level and a national forest inventory database. A REDD+ implementation plan for Central Sulawesi was also finalised in 2012.[20] Meanwhile, UNEP has been assisting Central Kalimantan with the development of a green growth strategy focusing on forestry and natural resources. Under this programme, the *2012 Indonesia Forest, Land and REDD+ Governance Index (Participatory Governance Assessment Report)* was also released, with the English version published on 25 June 2013 (see UN-REDD Programme 2013).

Donor bilateral arrangements

Another feature of public funding for REDD+ in Indonesia is the bilateral funding agreements with countries such as Australia and Norway. Under the 2008 Indonesia–Australia Forest Carbon Partnership (IAFCP), the Australian government committed AU$10 million to build capacity in Indonesia in anticipation of a REDD+ carbon market (see Irawan 2012). The Australian government also committed AU$30 million to establish the Kalimantan Forests and Climate Partnership. Under this Partnership, Australia and Indonesia were expecting to implement a demonstration activity in the carbon-rich peatland forests of Central Kalimantan. Key programmes under the IAFCP ended in June 2013, with targeted areas of work being extended to June 2014. The Australian government lists a number of activities that have been undertaken;[21] however, Australia has been criticised for withdrawing from the project before meeting the key goal of restoring 25,000 ha of

18 See www.un-redd.org/AboutUN-REDDProgramme/tabid/102613/Default.aspx.

19 To view the Resolution, see www.redd-monitor.org/2013/03/06/coonapip-panamas-indigenous-peoples-coordinating-body-withdraws-from-un-redd/#resolution.

20 Available at www.unredd.net/index.php?option=com_docman&task=doc_download&gid=8952&Itemid=53.

21 See http://aid.dfat.gov.au/countries/eastasia/indonesia/Pages/climate-change-init1.aspx.

peatland in Kalimantan. Previously the Australian government also shelved an AU$30 million REDD+ project in Sumatra (see Lang 2013).

In 2010 the governments of Norway and Indonesia signed a REDD+ agreement, known as the Climate Change Partnership, under which Norway provided Indonesia with US$1 billion to assist that country with the detailed and phased implementation of REDD+.[22] This Partnership is discussed in some detail here to demonstrate the influence that bilateral REDD+ agreements can have on individual tropical rainforest developing country jurisdictions. In the authors' view, it is a preferable model for national REDD+ capacity building compared with the unsuccessful and ad hoc REDD+ pilot project approach envisaged under the Indonesia–Australia Partnership. The Climate Change Partnership is also a specific example of the phased approach to REDD+ that is inherent in all of the international documentation on REDD+, as discussed in Chapter 2. The influence and success of the Climate Change Partnership in Indonesia is discussed at various points in this book. It is important to note, however, that, as a bilateral arrangement, the Partnership is not intended to be in conflict with any other REDD+ initiatives in Indonesia such as UN-REDD, the World Bank's Forest Carbon Partnership Facility or any other bilateral arrangements. The goals of the Partnership are to deliver significant GHG emissions reductions from deforestation, forest degradation and peatland conversion. At the same time, it should contribute to an international policy dialogue on REDD+ while supporting Indonesia's own REDD+ domestic programmes. It is clear that a key component of the Partnership is supposed to be the full and effective participation of all relevant stakeholders, including indigenous peoples, local communities and civil society, at all stages of implementation, although the authors note that there could be improvements in this regard. Financing by Norway depends on 'contributions-for-delivery' whereby payments will be made to Indonesia based on a progressive implementation of REDD+.

The intention of the Partnership is to implement the first two phases within 3–4 years of its establishment. The first phase of this Partnership is preparation in terms of which Indonesia was obliged to: develop a national REDD+ strategy to address the drivers of deforestation; establish a national agency reporting directly to the President and an independent agency for MRV progress on the REDD+ strategy; establish a funding instrument to be managed by a reputable funding institution; and select a State-wide pilot REDD+ programme.

The second phase, transformation, should have been implemented between 1 January 2011 and December 2013. In this phase, Indonesia must put in place the requirements for receiving 'contributions for verified emissions reduction'. This entails having in place a country-wide MRV system that is compliant with the international standards set by the Intergovernmental

---

22 To read the Letter of Intent, see www.unorcid.org/upload/doc_lib/Norway-Indonesia-LoI.pdf.

Panel on Climate Change[23] and which includes international verification of emissions reductions. Importantly, during this phase Indonesia had to: implement a two-year moratorium on all new concessions for peatland and forest conversion; establish a degraded lands database; enforce illegal logging laws and forest-related crimes and establish a special enforcement unit in this regard; and take appropriate measures to resolve land tenure conflicts and claims for compensation. Two State-wide pilot REDD+ projects should be established demonstrating that all aspects of phases one and two have been successfully delivered.

Phase three, 'contributions for verified emissions reduction', is to be implemented commencing in 2014. Here, Indonesia is to receive financial contributions from the Norwegian government for all emissions reductions that have progressed successfully through the MRV processes. These emissions reductions must be based on a previously established national reference level for Indonesia's greenhouse gas emissions from its forestry sector.

It is clear that this Climate Change Partnership has considerable potential for demonstrating a public funding model for REDD+ programmes in developing countries. It is also an example of how a developing country can move progressively through the three REDD+ stages to reach the point at which emissions reductions are fully verified and may subsequently be made available as commercially viable carbon credits in an international carbon credits market.

The state of the REDD+ carbon market

Despite the lack of an existing international legal framework for REDD+, an incipient market for REDD+ credits began emerging in many different jurisdictions from the mid-2000s. There were two principal reasons for this: the expectation of a strong compliance market and a burgeoning corporate interest in voluntarily going 'carbon neutral'.

REDD+ compliance markets

Prior to the Copenhagen negotiations, there was a high expectation that the UNFCCC negotiations for a post-2012 framework would deliver significant, legally binding emissions reduction commitments, at least for developed countries, and that this would translate to high demand for REDD+ offset credits. As this book shows, the climate change negotiations have incrementally developed a REDD+ framework to respond to this likely demand, at the same time as acknowledging the need to control emissions from deforestation in any case. In 2007 a number of countries began to prepare placing a cap on domestic emissions, with the United States, Australia and New Zealand pursuing policy and regulatory agendas to introduce emissions trading schemes,

---

23 Available at http://unfccc.int/resource/docs/2009/cop15/eng/11a01.pdf#page=11.

with a strong focus on REDD+ credits in the US.[24] However, there were at least two major interventions that slowed down the likelihood of a strong demand for REDD+ credits: weaker than expected outcomes at Copenhagen and the global financial crisis. Even though negotiations are currently underway for a legally binding emissions reduction agreement to come into effect in 2020, involving both developing and developed countries, the impetus that preceded Copenhagen has not yet been regained.

REDD+ *voluntary market*

Corporate and investor interest in the Voluntary Carbon Market (VCM) for REDD+ credits, in the absence of any internationally agreed standards, has given rise to the development of a wide variety of voluntary standards and projects. These include: the American Carbon Registry Forest Carbon Project Standard;[25] the Climate, Community and Biodiversity (CCB) Project Design Standards;[26] the Chicago Climate Exchange Standard; Plan Vivo Standards;[27] the Social Carbon Guidelines;[28] and the Verified Carbon Standard (VCS).[29] An analysis of the carbon offset standards shows that a complete and fully fledged carbon offset standard comprises the following three components: accounting standards; monitoring, verification and certification standards; and registration and enforcement standards (see Kollmuss et al. 2008). The accounting standards ensure that offsets are 'real, additional and permanent' while the monitoring, verification and certification standards ensure that offset projects perform against key project design criteria (Kollmus et al. 2008: 14). Verification and certification occur *ex post facto* and are distinguishable from validation of the project activity. Validation is usually undertaken by an independent auditor that reviews and validates the project design documents and other related documentation, such as environmental impact assessments and stakeholder consultation processes (Kollmus et al. 2008: 33). Finally, registration and enforcement systems ensure that offset credits are sold only once, with the registry containing publicly accessible information that identifies individual offset projects and tracks the ownership of offset credits. Contractual or legal standards should also stipulate the distribution

24 For a full discussion of this, see Lyster (2009).
25 Released in 2009: www.americancarbonregistry.org/carbon-accounting/ACR%20Forest%20 Carbon%20 Project%20Standard%20v1%20March%202009%20FINAL.pdf.
26 Released in 2008: www.climate-standards.org/ccb-standards/. Note that the Climate, Community and Biodiversity (CCB) Standards cannot be used on their own to generate carbon credits. Rather, the CCB Standards can be used to enhance other standards since they deal specifically with incorporating socioeconomic and environmental co-benefits in addition to consideration of carbon abatement.
27 Released in 2008: www.planvivo.org/wp-content/uploads/Plan-Vivo-Standards-20081.pdf.
28 Released in 2009, most recent version released July 2013: www.socialcarbon.org/wp-content/uploads/2012/11/SOCIALCARBON_STANDARD_v-5-.00.pdf.
29 Released in 2008, most recent version released 8 October 2013: www.v-c-s.org/sites/v-c-s.org/files/VCS%20Standard%2C%20v3.4.pdf.

of risk in the case of project failure, and perhaps also insurance arrangements for insuring against the risk.[30] Together, these aspects of the standards enable the trading of offset credits.[31]

An example of a voluntary REDD+ project in Indonesia is Merrill Lynch's signature of a six-year US$9 million agreement with Carbon Conservation (an Australian-based project developer) and UK-based NGO Fauna & Flora International (FFI) to buy voluntary emissions reductions (VERs) from a REDD+ project in Indonesia's Aceh province.[32] The Ulu Masen project area[33] comprises 750,000 ha with approximately 700,000 ha of this being contiguous with the Ulu Masen ecosystem. The forest areas include: 428,757 ha of unprotected State forest lands allocated for industrial scale logging, although this is not currently occurring; State forest lands zoned for commercial and community conversion logging licences; and 310,991 ha of State forest lands over which various categories of protected status have been proclaimed albeit that actual protection is weak and ineffective. The project time periods include a development phase (1 January 2008–31 December 2012) followed by a follow-up period for 25 years up to 2038.[34] In May 2011 East Asia Mineral Corporation acquired a 50 per cent equity interest in Carbon Conservation Pty Ltd. Despite Merrill Lynch's intention to invest US$9 million in carbon credits derived from the project none has yet been issued. Although the project was established in accordance with the CCB standards, the project has proved highly controversial[35] with a number of forestry non-government organisations criticising the project on various levels, including for not consulting adequately with indigenous people and local communities – a key concern of this book.

Another example of a VCM project in Indonesia is that located in Nagari Paninggahan located in the Singkarak watershed in West Sumatra province (see Neilson and Leimona 2013). *Nagari* is a territorial unit in Indonesia recognised as having its own political and judicial apparatus in which *adat* is acknowledged by specific *adat* regulations. This project was established in 2009 under two contractual arrangements between the Nagari Head and a private company, CO2 BV, representing various companies in the Netherlands. Under the first contract, 28 ha would be planted for agroforestry sequestering 4090 tonnes of CO_2. The contract price for this project was

---

30 To understand the way in which such risk is typically distributed under CDM offset projects, see the International Emissions Trading Association (IETA) Emissions Reduction Purchase Agreement (ERPA): www.ieta.org/assets/TradingDocs/cdmerpav.3.0final.doc.

31 *Ibid.*

32 However, recent media investigations conclude that nothing has come of this project in Aceh (Bachelard 2012).

33 See *Validation Audit Report for Nanggroe Aceh Darussalam, Flora and Fauna International, Carbon Conservation in Ulu Maasen Ecosystem (Aceh Province, Indonesia)*, available at http://www.rainforest-alliance.org/sites/default/files/climate_project/CarbonConservation_assessment.pdf.

34 *Ibid* at 11.

35 See, for example, http://www.redd-monitor.org/tag/ulu-masen/ (accessed 17 July 2014) and http://www.forestpeoples.org/sites/fpp/files/publication/2011/10/aceh-briefing-3.pdf.

US$29,473 and was paid in instalments depending on progress (Neilson and Leimona 2013: 224).

It is interesting that although there is no legislative basis in Indonesia for either of these Voluntary Carbon Market projects to proceed, they were nevertheless not excluded. In the *Paninggahan* case, for example, contracts were signed under a sophisticated pre-independence political system in West Sumatra whereby land tenure was governed by locally defined conventions and decision-making institutions. From the perspective of community members, this provided a legitimate basis for the acceptance of payments for the carbon emissions reductions derived from the project (Neilson and Leimona 2013: 224). These case studies demonstrate the need for the Indonesian government to gain control of the burgeoning voluntary market in Indonesia through the passing of national regulations for REDD+ projects. It is important to note, however, that the Indonesian experience is not unique. In 2008, for example, the REDD+ voluntary market under the Chicago Climate Exchange, moved ahead of any formal regulatory interventions, under the UNFCCC or by domestic governments, to invest in REDD+ credits especially in Latin America and Africa.[36]

A report on global Voluntary Carbon Markets in 2012 indicated that voluntary actors contracted 101 MtCO2e, which is 4 per cent more than in 2011 (although the market value decreased 11 per cent to US$523 million as offset prices fell slightly for several project types (see Peters-Stanley and Yin 2013)). Of this, 90 per cent of the demand for carbon offsets arose from the private sector, driven by corporate social responsibility and industry leadership. There was evidence of offset buyers' desire to positively impact the climate resilience of their supply chain or sphere of influence, indicating a strong relationship between buyers' business sectors and the project categories from which they contract offsets (Peters-Stanley and Yin 2013: v). Suppliers of voluntary carbon credits predict that market value could reach US$1.6–2.3 billion in 2020 (Peters-Stanley and Yin 2013: v).

As in 2011, REDD+ projects were buyers' third most popular choice as an offset source, transacting 6.8 MtCO2e or 8 per cent less than the previous year. This decline occurred exclusively in the categories of projects where a third-party standard to certify carbon reductions was not adopted or where a country-specific standard was utilised. However, in 2012 where REDD+ offsets were (or aim to be) certified to both the VCS and CCB Standards the transaction volume more than tripled. Clearly, buyers are more willing to support REDD+ projects that can verify climate, biodiversity and community benefits and then issue offsets (Peters-Stanley and Yin 2013: 21). In 2012 Indonesia was not the source of REDD+ voluntary carbon units (VCUs) as 96 per cent of offsets were issued from four VCS+CCB projects: the Kenyan Kasigau Corridor project (Phases I and II); the Mai Ndombe Project in the

Democratic Republic of the Congo (DRC); and the Alto Mayo Project in Peru (Peters-Stanley and Yin 2013: 21).

What this discussion indicates is that even while the COPs continue to negotiate the parameters of internationally sanctioned REDD+ activities in developing countries, the voluntary market has forged ahead. Voluntary standards have been developed and are being used to provide an assurance that the carbon, and biodiversity, benefits are being delivered at the project level.

Conclusion

This chapter has sought to identify a number of key concerns regarding REDD+. The REDD+ international documentation, arising out of the COPs under the UNFCCC, appears to gloss over the many substantial challenges associated with implementing credible and legitimate REDD+ schemes. This chapter also demonstrates one of the key REDD+ complexities that relates to how REDD+ activities in developing countries ought to be funded. Should REDD+ activities generate carbon credits for compliance and/or voluntary markets and, if so, can they be relied on as genuine emissions reduction credits? Or should REDD+ activities not form part of the international emissions trading regime at all and rather be publicly funded? As Chapter 2 demonstrates, the COPs have moved, and continue to move, in an incremental fashion to address many of these issues.

References

Avitabile, V. (2013) 'Measuring tropical forest carbon stocks', in R. Lyster, C. MacKenzie and C. McDermott (eds) *Law, tropical forests and carbon: The case of REDD+*, New York: Cambridge University Press.

Bachelard, M. (2012) 'Credits lost in tangle of Aceh's forest', *Sydney Morning Herald*, 9 June: www.smh.com.au/environment/conservation/credits-lost-in-tangle-of-acehs-forest-20120608-201gl.html.

Bond, I., M. Grieg-Gran, S. Wertz-Kanounnikoff, P. Hazlewood, S. Wunder and A. Angelsen (2009) *Incentives to sustain forest ecosystem services: A review and lessons for REDD*, Natural Resouce Issues No. 16, London: International Institute for Environment and Development, with CIFOR, Bogor, Indonesia, and World Resources Institute: www.iied.org/pubs/pdfs/13555IIED.pdf.

David C., F. Daviet, S. Nakhooda and A. Thuault (2009) *A review of 25 Readiness Plan Idea Notes from the World Bank Forest Carbon Partnership Facility*, Working Paper, Washington, DC: World Resources Institute, February: www.redd-monitor.org/wordpress/wp-content/uploads/2009/03/r-pin_analysis.pdf.

Fisher, R. and R. Lyster (2013) 'Land and resource tenure: The rights of indigenous peoples and forest dwellers', in R. Lyster, C. MacKenzie and C. McDermott (eds) *Law, tropical forests and carbon: The case of REDD+*, New York: Cambridge University Press.

Forest Carbon Partnership Facility (2013) *A guide to the FCPF Readiness Assessment Framework (June 2013)*, Washington, DC: Carbon Finance, World Bank:

www.forest carbonpartnership.org/sites/fcp/files/2013/July2013/FCPC%20frame work%20 text%207-25-13%20ENG%20web.pdf.

Griffiths, T. (2007) 'Seeing "RED?" Avoided deforestation and the rights of Indigenous peoples and local communities', Forest Peoples Programme.

Heal, G., G. C. Daily, P. R. Ehrlich, J. Salzman, C. Boggs, J. Hellmann et al. (2001) 'Protecting natural capital through ecosystem service districts', *Stanford Environmental Law Journal* 30: 333.

Irawan, S. (2012) *Provincial REDD+ implementation plan: Central Sulawesi*, Jakarta, Indonesia: UN-REDD Programme Indonesia: www.climatechange.gov.au/ international/ publications/pubs/indonesia-australia.pdf.

Karousakis, K. (2006) 'Initial review of policies and incentives to reduce GHG emissions from deforestation', Organisation for Economic Co-operation and Development (OECD), October.

Karousakis, K. and J. Corfee-Morlot (2007) 'Financing mechanisms to reduce emissions from deforestation: Issues in design and implementation', OECD and International Energy Agency (IEA), December.

Kollmuss, A., H. Zink and C. Polycarp (2008) *Making sense of the voluntary carbon market: A comparison of Carbon Offset Standards*, WWF, March: http://assets.panda. org/downloads/vcm_report_final.pdf.

Lang, C. (2013) 'Australia shuts down the Kalimantan Forest Carbon Partnership: "A lot of funds spent and very little progress"', REDD-monitor, 4 July: http:// www.redd-monitor.org/2013/07/04/australia-shuts-down-the-kalimantan-forest-carbon-partnership-a-lot-of-funds-spent-and-very-little-progress/.

Lyster, R. (2009) 'The new frontier of climate law: Reducing emissions from deforestation (and degradation)', *Environmental and Planning Law Journal* 26(2): 417.

Lyster, R. (2012) '(De)regulating the rural environment', *Environmental Planning and Law Journal* 19(1): 34.

McDermott, C. (2013) 'REDD+ and multi-level governance: Governing for what and for whom?', in R. Lyster, C. MacKenzie and C. McDermott (eds) *Law, tropical forests and carbon: The case of REDD+*, New York: Cambridge University Press.

Meridian Institute (2009) *Reducing Emissions from Deforestation and Forest Degradation (REDD): An options assessment report*, Prepared for the Government of Norway, March: www.redd-oar.org/.

Neilson, J. and B. Leimona (2013) 'PES and environmental governance in Indonesia' in R. Lyster, C. MacKenzie and C. McDermott (eds) *Law, tropical forests and carbon: The case of REDD+*, New York: Cambridge University Press.

Peskett, L., D. Brown and C. Luttrell (2006) 'Can payments for avoided deforestation to tackle climate change also benefit the poor?', Forest Briefing Paper No. 12, London: Overseas Development Institute, November.

Peters-Stanley, M. and D. Yin (2013) *Maneuvering the mosaic: State and trends of the Voluntary Carbon Market 2013*, Ecosystem Market Place and Bloomberg New Energy Finance, 20 June: www.forest-trends.org/documents/files/doc_3898.pdf.

Sari, A. P., M. Maulidya, R. N. Bhutarbar, R. E. Sari and W. Rusmantoro (2007) *Executive summary: Indonesia and climate change*, March.

UN-REDD Programme (2013) 'Joining forces to tackle governance challenges on land, forest and REDD+ in Indonesia: Launch of Governance Forestry Index', media release, Lombok, Indonesia, 25 June: www.un-redd.org/PGAPressRelease/ tabid/128492/ Default.aspx.

2 The international legal framework for REDD+

Prior to the 2007 climate change negotiations at the Thirteenth Conference of the Parties (COP, specifically COP13) to the United Nations Framework Convention on Climate Change (UNFCCC)[1] in Bali, REDD+ was not incorporated in the international legal framework for addressing climate change. Since Bali there have been five significant negotiations: COP15 in Copenhagen in December 2009; COP16 in Cancun in December 2010; COP17 in Durban in December 2011; COP18 in Doha in November 2012; and COP19 in Warsaw in November 2013. Each of these summits has delivered outcomes that have significant ramifications for achieving the goal of keeping global temperature rise below 2°C above pre-industrial temperatures and for REDD+ in contributing to this objective. This chapter analyses the outcomes of these COPs with a particular focus on REDD+ and raises key legal questions regarding the implementation of the REDD+ safeguards.[2]

The Bali Action Plan

Under the Bali Action Plan, agreed at COP13, the Parties decided to begin a process towards reaching a decision at COP15 on a shared vision for long-term cooperative action on climate change. This vision would include a long-term global goal for emissions reductions, based on the principle of common but differentiated responsibilities.

Significantly, the Action Plan required 'enhanced consideration of policy approaches and positive incentives on issues relating to reducing emissions from deforestation and forest degradation in developing countries; and the role of conservation, sustainable management of forests and enhancement of forest carbon stocks in developing countries' (Article 1(b)(iii)).[3] This builds on a decision taken at COP11, in December 2005, to establish a two-year

---

1 *United Nations Framework Convention on Climate Change*, opened for signature 20 June 1992, 31 ILM 848 (1992) (entered into force 21 March 1994).
2 This chapter builds on a previous publication (Lyster 2011). See also Lyster (2009); Fisher and Lyster (2013).
3 Available at: http://unfccc.int/files/meetings/cop_13/application/pdf/cp_bali_action.pdf.

review of relevant scientific and methodological issues and to consider policy approaches and incentives for reducing emissions from deforestation in developing countries.[4]

The incorporation of REDD+ in the Bali Action Plan was highly significant as it received little or no attention in prior negotiations. Under Article 3.3 of the Kyoto Protocol,[5] Annex I Parties[6] may rely on *domestic* reductions in global greenhouse gas (GHG) emissions resulting from forestry activities, limited to afforestation[7] and reforestation[8] since 1990, to meet their emissions reduction targets under the Protocol.[9] Similarly, afforestation and reforestation projects undertaken by Annex I Parties in developing countries may be relied on, under the Clean Development Mechanism (CDM),[10] to satisfy their Kyoto commitments.

There are a number of reasons why REDD+ was excluded from the provisions of the Kyoto Protocol, including from the CDM. These include concerns about: the risk of leakage;[11] non-permanence;[12] establishing baselines;[13] additionality;[14] and difficulties associated with monitoring and measurement. During the two-year review of REDD+ established at COP11, considerable advances were made in addressing these problems, particularly with respect

4 As part of this process, a number of workshops were organised under the auspices of the UNFCCC Subsidiary Body of Scientific and Technical Advice, including in Italy in September 2006, Australia in March 2007, Bonn in May 2007 and Bali in December 2007: see, for example, SBSTA (2007).

5 *Kyoto Protocol to the Framework Convention on Climate Change*, opened for signature 16 March 1998, 37 ILM 22 (1998) (entered into force 16 February 2005).

6 These are developed countries with emissions reduction targets under the Kyoto Protocol.

7 'Afforestation' is defined as the direct human-induced conversion of land that has not been forested, for a period of at least 50 years, to forested land through planting, seeding and/or the human-induced promotion of natural seed sources (see UNFCCC 2002: 54).

8 'Reforestation' is defined as the direct human-induced conversion of non-forested land to forested land through planting, seeding and/or the human-induced promotion of natural seed sources, on land that was forested but that has been converted to non-forested land. For the first commitment period, reforestation activities will be limited to reforestation occurring on those lands that did not contain forest on 31 December 1989 (see UNFCCC 2002: 54).

9 Although reliance on this is limited in accordance with the Marrakesh Accords, negotiated at COP7 in 2001; see Decision 11/CP.7: http://unfccc.int/files/meetings/workshops/other_meetings/application/pdf/ 11cp7.pdf.

10 The CDM (Article 12 of the Kyoto Protocol) allows developed countries to invest in emission-reducing projects in developing countries and to obtain certified emission reduction units (CERs) towards meeting their assigned amounts.

11 'Leakage' refers to greenhouse gas emissions that occur outside the project boundary, but that are nevertheless attributable to its activities.

12 'Permanence' refers to the possibility that carbon is released into the atmosphere as a result of fire, illegal logging or a change in government.

13 Credits can only be generated for emissions below the 'baseline', i.e. GHG emissions reduction that would not have occurred in the absence of a CDM project.

14 It must be demonstrated that the carbon sequestration would not have occurred without the incentives provided by the project.

to monitoring and measurement.[15] It should also be noted that many of the problems relating to REDD+ also arise with respect to afforestation and reforestation CDM projects. For this reason, the CDM Executive Board has developed unique rules governing these types of projects, although, in practice, they have not eventuated given the temporary nature of the credits issued under the rules.[16]

Outcomes of the UNFCCC negotiations since 2009: a focus on REDD+

In this section, we survey the principal outcomes of the Copenhagen (2009), Cancun (2010), Durban (2011), Doha (2012) and Warsaw (2013) Conference of the Parties to the UNFCCC. Our focus is ultimately the REDD+ outcomes at each of these COPs. As will become clear, the REDD+ programme of action has been an iterative and slowly evolving one, as each COP has made incremental additions to the implementation of REDD+ programmes in developing countries.

Although the Copenhagen Accord[17] is not a legally binding treaty, it had a major impact on international climate policy and for this reason its acknowledgment of the crucial role of REDD+ is highly significant. Parties to the Accord agreed also on the need to mobilise financial resources from developed countries for REDD+ (UNFCCC 2009: [6]).[18] Given the failure of the Copenhagen negotiations to deliver a legally binding agreement, the Accord has been subsumed under the Cancun Agreements negotiated at COP16. The Cancun Agreements were further refined at COP17. For this reason, these Agreements will be discussed rather than the Accord, while their provisions for REDD+ will be analysed separately.

Since COP17 built on the Cancun Agreements[19] negotiated at COP16 in December 2010, which confirmed the parameters of the December 2009

15 See, for example, the 2006 Intergovernmental Panel on Climate Change (IPCC) Guidelines for National Greenhouse Gas Inventories – Agriculture, Forestry and Other Land Use: www.ipcc-nggip.iges.or.jp/public/2006gl/vol4.htm; and the Global Observation for Forest Cover and Land Dynamics (GOFG-GOLD) REDD Sourcebook, which uses remote sensing to monitor and measure greenhouse gas emissions from forests: www.gofc-gold.uni-jena.de/redd/.

16 See http://cdm.unfccc.int/methodologies/ARmethodologies/approved_ar.html; also Streck and Scholz (2006: 868).

17 Available at UNFCCC (2009).

18 It is also important to note the REDD+ Partnership developed under the Oslo-Paris Accord, which following its meeting on 27 May 2010 has committed additional REDD+ funds; see REDD+ Partnership document: www.oslocfc2010.no/pop.cfm?FuseAction=Doc&pAction=View&pDocumentId=25019.

19 See 'The Cancun Agreements: 2010 Outcome of the work of the Ad Hoc Working Group on Long-term Cooperative Action under the Convention' (Decision 1/CP.16), 'The Cancun Agreements: Outcome of the work of the Ad Hoc Working Group on Further Commitments for Annex I Parties under the Kyoto Protocol at its fifteenth session' (Decision 1/CMP.6) and 'The Cancun Agreements: Land use, land-use change and forestry' (Decision 2/CMP.6): UNFCCC (2011: 2, 3, 5 respectively).

Copenhagen Accord (UNFCCC 2010), it is necessary to refer to both COPs. Both COP16 and COP17 proceeded along the two-track process established under the Bali Action Plan adopted at COP13 in 2007. The two-track process comprised the Ad Hoc Working Group on Long-term Cooperative Action under the UNFCCC (AWG-LCA) and the Ad Hoc Working Group on Further Commitments for Annex I Parties under the Kyoto Protocol (AWG-KP). These two working groups were established to reach a decision on whether future legally binding GHG reduction commitments would be made under the Kyoto Protocol. They met quarterly to conduct further negotiations following the Bali summit. At COP16, the AWG-LCA text, which included the essential elements of the Copenhagen Accord, was adopted by the COP and decisions made in that text are now binding on the COP under the UNFCCC. For the purposes of this chapter, the decisions on adaptation, technology and capacity building will not be referred to.

Global average temperatures

The COP17 AWG-LCA text[20] builds on agreements reached at COP16. Both COP16 and 17 recognised the need to hold the increase in global average temperatures to *below* 2°C above pre-industrial levels, as well as the need to consider strengthening the long-term goal in relation to a global average temperature rise of 1.5°C on the basis of the best available scientific knowledge (Draft decision [-/CP.17], Article 4).[21] At both COP16 and COP17, the parties agreed that they should cooperate in achieving the peaking of GHG emissions as soon as possible (Decision 1/CP.16, UNFCCC 2011: 2, Article 6).[22]

Extending the life of the Kyoto Protocol

COP17 decided, under the auspices of the AWG-KP (Draft decision -/CMP.7, Article 5),[23] that a second commitment period of the Protocol would begin on 1 January 2013 and end either on 31 December 2017 or 31 December 2020 (subject to a decision of the AWG-KP). Annex I Parties were invited to convert their Kyoto Protocol targets to 'quantified emission limitation or reduction objectives' (QELROs) for the second commitment period and to

20 See Draft decision [-/CP.17]: *Outcome of the work of the Ad Hoc Working Group on Long-term Cooperative Action under the Convention* (2011): http://unfccc.int/files/meetings/durban_nov_2011/decisions/application/pdf/cop17_lcaoutcome.pdf.

21 See also *Establishment of an Ad Hoc Working Group on the Durban Platform for Enhanced Action*: http://unfccc.int/files/meetings/durban_nov_2011/decisions/application/pdf/cop17_durban-platform.pdf.

22 See also Draft decision [-/CP.17], Article 2: http://unfccc.int/files/meetings/durban_nov_2011/decisions/ application/pdf/cop17_lcaoutcome.pdf.

23 See Draft decision -/CMP.7: *Outcome of the work of the Ad Hoc Working Group on Further Commitments for Annex I Parties under the Kyoto Protocol at its Sixteenth session* (2011): http://unfccc.int/files/meetings/durban_nov_2011/decisions/application/pdf/awgkp_outcome.pdf.

submit information on their intentions to the AWG-KP by 1 May 2012. A seventh GHG, nitrogen trifluoride (NF_3), was added to Annex A of the Protocol (Draft decision -/CMP.7, Annex II).

Developing country commitments

It is well known that developing countries made no commitments to emissions reductions under the Kyoto Protocol, consistently with the principle of common but differentiated responsibility. While the principle of common but differentiated responsibility is still alive and well, there has been a distinct shift with regard to negotiations for a post-Kyoto world. While COP17 recognises that social and economic development and poverty eradication are the first and overriding priorities of developing economies (Draft decision [-/CP.17], Article 87), and that the share of global emissions originating in developing countries will grow to meet their social and developmental needs, a low-emissions development strategy is nevertheless regarded as central to sustainable development (Draft decision [-/CP.17], Part B, Preamble). COP17 notes the voluntary commitments made by developing countries to prepare and implement nationally appropriate mitigation actions (NAMAs) under the Cancun Agreements (Part B, Preamble).[24] REDD+ programmes are recognised as legitimate NAMA activities. Developing countries that had not yet made commitments were encouraged to do so (Draft decision [-/CP.17], Article 32). COP17 also decided to hold workshops in May 2012 to further understand these commitments (Article 33) and invited developing countries to submit more information on their NAMAs to the Secretariat by 5 March 2012 (Article 35). A web-based registry is to be developed, from which developing countries can seek international support for their NAMAs (Article 45).

Developed countries are invited to submit information to the Secretariat on financial, technical and capacity-building support, which they can offer for the preparation and implementation of NAMAs (Draft decision [-/CP.17], Article 48). The registry will facilitate the matching of actions seeking international support with support that is available (Article 51).

It was decided at COP16 that internationally supported NAMAs will be subject to domestic monitoring, reporting and verification (MRV) procedures, conducted internally by the relevant country, and also be subject to international MRV in accordance with guidelines to be developed under the Conventions Decision 1/CP.16, UNFCCC 2011: 2, Article 61. NAMAs which are only domestically supported will be subject to domestic MRV in accordance with general guidelines to be developed under the UNFCCC (Decision 1/CP.16, Article 62). COP17 requests the development of guidelines for domestic MRV of domestically supported NAMAs (Draft decision [-/CP.17], Article 37).

24 Developing countries' voluntary commitments can be viewed at: http://unfccc.int/resource/docs/2011/ awglca14/eng/inf01.pdf.

Developing a new legally binding protocol, instrument or outcome

It is important to note that, at COP17, it was decided to terminate the AWG-LCA established in Bali at the end of 2012. It was decided instead that a new body, the Ad Hoc Working Group on the Durban Platform for Enhanced Action (ADP),[25] would be established, noting with 'grave concern' that the significant gap between the Parties' current mitigation pledges by 2020 are unlikely to hold the increase in global average temperatures below 2°C, or 1.5°C, above pre-industrial temperatures.[26] This Working Group started its work as a matter of urgency in the first half of 2012 and will complete its work as early as possible, but no later than 2015, to adopt 'a protocol, or other legal instrument or an agreed outcome with legal force under the UNFCCC'. The Protocol must be adopted at COP21 in 2015 and come into effect from 2020. The process must raise the level of ambition of all Parties, including all developed and developing countries, and must be informed by the IPCC's Fifth Assessment Report and the outcomes of the progress review to be conducted between 2013–15.[27] A work plan will be launched to close the ambition gap, to ensure the highest possible mitigation efforts by all Parties.

The Eighteenth Conference of the Parties (COP18) to the UNFCCC, also known as the Doha Climate Gateway, and the Eighth Meeting of the Parties (MOP) to the Kyoto Protocol proceeded across three tracks:

- the Advance Working Group on Long-term Cooperative Action (AWG-LCA)
- the Advance Working Group on Further Commitments for Annex I Parties (developed countries) under the Kyoto Protocol (AWG-KP)
- the Ad Hoc Working Group on the Durban Platform for Enhanced Action (ADP).

However, at COP18 the work of both the AWG-LCA and the AWG-KP concluded, so all future international negotiations must now occur under the auspices of the ADP. A number of decisions were made at Doha (see Draft decision -/CMP.8, UNFCCC 2012), most notably the decision that the second commitment period of the Kyoto Protocol will run between 1 January 2013 and 31 December 2020 (UNFCCC 2012, Article 4). Furthermore, developed countries should achieve aggregate emissions reductions of 25 to 40 per cent below 1990 levels by 2020 (Article 7). It is important to note, however, that Canada withdrew from the Protocol in November 2012, the United States has never ratified the treaty and Russia and Japan have not signed up

---

25 See *Establishment of an Ad Hoc Working Group on the Durban Platform for Enhanced Action*: http://unfccc.int/files/meetings/durban_nov_2011/decisions/application/pdf/cop17_durbanplatform.pdf.

26 *Ibid.*, Preamble.

27 *Ibid.*, Article 6.

to the second commitment period. This compromises the overall reductions expected of developed economies. Essentially, it will be the task of the ADP to work with developed and developing countries to achieve the level of reductions needed to reach all of the Parties' stated goal:[28] deep reductions in GHG emissions to hold the increase in global average temperatures to *below* 2°C above pre-industrial leve_s, in accordance with the principle of common but differentiated responsibility, and to attain a global peaking of emissions as soon as possible (Draft decision -/CMP.8, UNFCCC 2012, Article 1).

The financial mechanism

The financial mechanism adopted at both COP16 and COP17 is obviously relevant to REDD+ given the level of financial support needed by tropical rainforest countries to adopt REDD+ programmes. Suffice it to say that COP16 took note of the collective commitment by developed countries made at Copenhagen to provide US$30 billion in fast-start funding for the period 2010–12 (Decision 1/CP.16, UNFCCC 2011: 2, Article 95). In the interests of transparency, developed countries were invited to report in May 2011, 2012 and 2013 to the UNFCCC Secretariat on the resources they have provided (Decision 1/CP.16, Article 96). With regard to long-term finance, COP16 decided in accordance with the UNFCCC that scaled-up, new, predictable and adequate funding must be provided to developing country parties, taking into account the urgent and immediate needs of developing countries that are particularly vulnerable to the adverse effects of climate change (Article 97). Developed countries will commit US$100 billion per year by 2020 to address these needs (Article 98), with the funds coming from public and private, bilateral and multilateral and alternative sources (Article 99). The COP also established a Standing Committee on Finance to assist the COP in exercising its functions under the Financial mechanism. Subsequently, COP17 decided to adopt various procedures to assist the COP to improve coherence and coordination in the delivery of climate financing, the mobilisation of financial resources and MRV of support provided to developing countries.[29] The composition and modalities of the Standing Committee are also stipulated (Article 47).

Reducing Emissions from Deforestation and Degradation (REDD+)

COPs 16 and 17 affirmed that developing countries should collectively aim to slow, halt and reverse forest cover and carbon loss according to national circumstances and find effective ways to reduce human pressure on forests that

---

28 Draft decision -/CP.18: http://unfccc.int/2860.php#decisions.
29 Draft decision [-/CP.17], Artic_e 121: http://unfccc.int/files/meetings/durban_nov_2011/ decisions/ application/pdf/cop17_lcaoutcome.pdf.

results in GHG emissions – including by addressing drivers of deforestation. COP16 encouraged countries to reduce emissions from deforestation and forest degradation; conserve forest carbon stocks; manage forests sustainably; and enhance forest carbon stocks, known collectively now as REDD+ (the REDD+ activities).[30] Extensive provisions on REDD+ can be found in Annexes I and II of the AWG-LCA's Cancun Agreements, as discussed later. Likewise, at COPs 18 and 19 the REDD+ provisions were significantly advanced and enhanced. Any reading of these provisions is necessarily highly detailed and complex given that they impose legally binding obligations on countries. However, the authors draw attention to the ways in which the obligations are qualified by use of the words 'as appropriate', 'in accordance with national circumstances', and 'in accordance with the availability of financial resources'. Such qualifications cast considerable doubt on whether the obligations set out in the texts are indeed enforceable.

Overview of REDD+ activities and responsibilities

At COP16, in order to support the agreed REDD+ activities, REDD+ countries were requested to develop:

- a national strategy or action plan;
- a national forest reference emission level and/or forest reference level (these should be a combination of sub-national forest reference emissions levels and/or forest reference levels);
- a robust and transparent national forest monitoring system for monitoring and reporting on activities with, if appropriate, subnational monitoring and reporting as an interim measure;
- a system for providing information on how the REDD+ safeguards (see below) are being addressed and respected (Decision 1/CP.16, Article 71).

Countries were also requested to address the drivers of deforestation and forest degradation, land tenure and forest governance issues, gender considerations and the REDD+ safeguards, ensuring the full and effective participation of relevant stakeholders including indigenous peoples and local communities. The COP decided that REDD+ activities will be implemented in phases beginning with the development of national strategies or action plans, policies and measures and capacity building; followed by their implementation that could involve further capacity building, technology development and transfer (Decision 1/CP.16, clause 2). Meanwhile, demonstration activities should be

---

30 'The Cancun Agreements: 2010 Outcome of the work of the Ad Hoc Working Group on Long-term Cooperative Action under the Convention' (Decision 1/CP.16), Article 70: UNFCCC (2011: 2). Note that REDD+ extends beyond deforestation and forest degradation, to also be concerned with the role of conservation, sustainable management of forests and enhancement of forest carbon stocks.

fully measured, reported and verified (Article 73). Developed countries were urged to support the REDD+ activities of developing countries.

Guidance and safeguards for REDD+

Appendix I to the COP16 AWG-LCA text, which provides guidance and safeguards for REDD+, requires these activities to:

- be country driven and consistent with the objective of environmental integrity, and take into account the multiple functions of forests and ecosystems;
- be undertaken in accordance with national development priorities, objects and circumstances and respect national sovereignty while being consistent with sustainable development needs and goals;
- be implemented in the context of sustainable development and reducing poverty, while responding to climate change and being consistent with the adaptation needs of the country;
- be supported by adequate and predictable financial and technology support, including support for capacity building, and be results based and promote sustainable management of forests (Decision 1/CP.16, Appendix I, clause 1).

Crucially, with regard to the legitimacy of REDD+ programmes, the safeguards emanating from COP16 include that: actions must complement or be consistent with the objectives of national forest programmes and relevant international conventions and agreements; transparent and effective national forest governance structures must be promoted and supported in accordance with national legislation and sovereignty, while respecting the knowledge and rights of indigenous peoples and members of local communities. Appendix I notes that the United Nations (UN) General Assembly has adopted the UN Declaration on the Rights of Indigenous Peoples[31] and that the full and effective participation of relevant stakeholders, in particular indigenous peoples and local communities, must be promoted and supported. Actions must also be consistent with the conservation of natural forests and biological diversity and must address the risk of reversals and reduce the displacement of emissions (Decision 1/CP.16, UNFCCC 2011: 2, clause 2).

COP17 recalls the REDD+ agreements at COP16 regarding safeguards for indigenous people and local communities, results-based actions following a staged development of REDD+ programmes and that these actions should be subject to monitoring, reporting and verification (Draft decision [-/CP.17], Article 64).[32]

---

31 *United Nations Declaration on the Rights of Indigenous Peoples*, opened for signature 13 September 2007, UN Doc. A/RES/61/295 (2007) (entered into force 2 October 2007).

32 Note also that a separate decision was made on safeguards and reference levels: http://unfccc. int/files/meetings/durban_nov_2011/decisions/application/pdf/cop17_safeguards.pdf.

COP17 adopted the advice of the Subsidiary Body for Scientific and Technological Advice (SBSTA), requested at COP16, and produced a separate REDD+ Guidance document.[33]

The Guidance notes that implementation of the COP16 safeguards, and information on how these safeguards are being addressed and respected, should be included in (where appropriate) all phases of REDD+ implementation.[34] The inclusion of the words 'where appropriate' has caused concern among some non-government organisations (NGOs) given that adherence to the safeguards was not qualified in this way at COP16.[35] Furthermore, COP17 qualified the COP16 imperative for developing countries to provide information on how the safeguards are addressed and respected by stating that 'national circumstances and respective capabilities' should be taken into account while 'recognising national sovereignty and legislation'. Although relevant international obligations and agreements and gender considerations should also be taken into account, COP17 should be regarded as watering down the requirement to provide information as established at COP16. The Guidance goes on to state that the information provided should: be transparent, consistent and accessible by all relevant stakeholders and updated on a regular basis; be flexible to allow for improvements over time; be country driven and implemented at the national level; and build on existing systems, as appropriate (Decision 1/CP.16, UNFCCC 2011, clause 2). Again, the reference to 'flexibility' and 'allow[ing] for improvements', with regard to the provision of information regarding the safeguards, modifies the requirements of COP16.

Reporting on REDD+ safeguards

It was also agreed at COP17 that developing country Parties undertaking REDD+ activities should provide a summary of information on how all of the safeguards are being addressed and respected throughout the implementation of these activities (Decision 1/CP.16, UNFCCC 2011, clause 3). This should be provided periodically and be included in the countries' national communications (UNFCCC 2011, clause 4). The COP requested the SBSTA, at its meeting in May 2012, to consider the timing of the first presentation (and the frequency of subsequent presentations) of the summary of information, with a view to recommending the adoption of a decision on this at COP18 in December 2012 (clause 5). It also requested the SBSTA to consider and report to COP18 on the need for further guidance to ensure transparency, consistency, comprehensiveness and effectiveness when informing on how all safeguards are addressed and respected (clause 6).

33 *Guidance on systems for providing information on how safeguards are addressed and respects and modalities relating to forest reference emissions levels and forest reference levels as referred to in decision 1/CP.16*: http://unfccc.int/files/meetings/durban_nov_2011/decisions/application/ pdf/cop17_safeguards. pdf. The SBSTA recommendation is also available (SBSTA 2011).
34 *Ibid.*, clause 1.
35 See Dooley and Horner (2012).

At COP19, a decision on the requirement for REDD+ countries to present information on the REDD+ safeguards was subsequently taken.[36] It may come as a surprise to many that the COP decided that, after beginning the implementation of REDD+ activities, developing country Parties should start providing the safeguards information in their national communications, but that they could also do so voluntarily on the web platform on the UNFCCC website (Decision 12/CP.19, Article 4) (as discussed later). The COP also decided that the frequency of reporting thereafter should be consistent with the national communications requirements for non-Annex I countries, but on a voluntary basis on the web platform (Article 5). The authors regard this as a significant concession on the part of the COP to REDD+ countries that might otherwise be reluctant to report formally in their national communications on their implementation of the REDD+ safeguards. This does not augur well for the future implementation of the safeguards.

The drivers of deforestation and modalities for establishing national forest emissions levels and measuring, reporting and verification (MRV)

Appendix II to the COP16 AWG-LCA text requested the SBSTA to: identify the drivers of deforestation and forest degradation in developing countries; and to identify the associated methodological issues to estimate emissions and removals from these activities; assess their potential contribution to the mitigation of climate change; and report to COP18 on these matters.[37] SBSTA was also requested to develop modalities relating to establishing national forest reference emissions levels (Decision 1/CP.16, Appendix II, clause (b)) and develop the necessary modalities for measuring, reporting and verifying anthropogenic forest-related emissions by sources and removals by sinks (Appendix II, clause (b)).

The drivers of deforestation

At COP19 in Warsaw, the drivers of deforestation and forest degradation were addressed. The COP noted the complexity and multiple drivers of deforestation and forest degradation, the different national circumstances and that livelihoods may be dependent on activities relating to these drivers.[38]

---

36 Decision 12/CP.19: *The timing and the frequency of presentations of the summary of information on how all the safeguards referred to in decision 1/CP.16, appendix I, are being addressed and respected*: http://unfccc.int/files/meetings/warsaw_nov_2013/decisions/application/pdf/cop19_safeguards_1cp16a1.pdf.
37 Decision 1/CP.16, Appendix II, clause (a): UNFCCC (2011: 2).
38 See Decision 10/CP.19: *Addressing the drivers of deforestation and forest degradation*: http://unfccc.int/files/meetings/warsaw_nov_2013/decisions/application/pdf/cop19_drivers_deforestation.pdf.

The COP reaffirmed the importance of addressing these drivers in the context of REDD+ national strategies and action plans (Decision 15/CP.19, Article 1). The COP also recognised that the drivers have many causes and that actions to address them would be unique according to the national circumstances and capabilities of developing countries (Article 2). The COP encouraged all Parties, organisations and the private sector and other stakeholders to take action to reduce and address these drivers, while also sharing the results of their work on the UNFCCC REDD+ web platform (Articles 3 and 4). Developing countries were encouraged to take note of this information (Article 5).

Modalities for forest reference emission levels and forest reference levels

COP17 provided a REDD+ *Guidance* which included the modalities for forest reference emission levels and forest reference levels.[39] Here, COP17 agreed with the COP16 decision that forest reference emission levels and/or forest reference levels, expressed in tonnes of carbon dioxide equivalent (CO_{2e}) per year, are benchmarks for assessing each country's performance in implementing REDD+ activities. It decided that forest reference emission levels and forest reference levels should be established transparently, taking into account historic data and adjusting for national circumstances (Decision 12/CP.17, clause 7).[40] They should also be consistent with anthropogenic forest-related greenhouse gas emissions by sources and removals by sinks, as contained in each country's greenhouse gas inventories (clause 8). The Parties are invited to submit information on, and the rationale for, the development of their forest reference emission levels and/or forest reference levels – including details of national circumstances and, if adjusted, details on how the national circumstances were considered (clause 9).

COP17 also regarded a step-wise approach to the national forest reference emission level and/or forest reference level development as useful, enabling Parties to gradually improve the forest reference emission level and/or forest reference level by incorporating better data, improved methodologies and, where appropriate, additional pools. This can only be achieved in the context of the provision of adequate and predictable financial, technical and technological support given to REDD+ countries (Decision 12/CP.17, clause 10). It is acknowledged that subnational forest reference emission levels and/or forest reference levels may be elaborated as an interim measure, while

39 Forest reference emission levels and/or forest reference levels expressed in tonnes of CO_{2e} per year are benchmarks for assessing each country's performance in implementing its REDD+ activities; see Decision 12/CP.17: *Guidance on systems for providing information on how safeguards are addressed and respected and modalities relating to forest reference emission levels and forest reference levels as referred to in decision 1/CP.16*, 16 [7]: http://unfccc.int/resource/docs/2011/cop17/eng/09a02.pdf# page=16.

40 Here the COP refers to Decision 4/CP.15, [7] reached at the Copenhagen COP (UNFCCC 2010: 11).

transitioning to a national forest reference emission level and/or forest reference level (clause 11). It is also agreed that a REDD+ country should update its forest reference emission level and/or forest reference level periodically as appropriate, taking into account new knowledge, new trends and any modification of scope and methodologies (clause 12). REDD+ countries are invited, on a voluntary basis and when deemed appropriate, to submit proposed forest reference emission levels and/or forest reference levels to the COP (clause 13) and the UNFCCC Secretariat is requested to make this information available on the UNFCCC REDD+ web platform (clause 14). The COP agreed to establish a process that enables technical assessment of the proposed forest reference emission levels and/or forest reference levels when submitted or updated by Parties (clause 15).

Guidelines are also provided to REDD+ countries for the provision of information on their reference levels (Decision 12/CP.17, Annex). Each country should include in its submission information that is transparent, complete and consistent with the COP guidance. It must also be accurate so as to allow a technical assessment of the data, methodologies and procedures used in the construction of a forest reference emission level and/or forest reference level. The information provided should be guided, as is appropriate, by the most recent IPCC guidance and guidelines, as adopted or encouraged by the COP. The information should include:

- information that was used by Parties in constructing a forest reference emission level and/or forest reference level, including historical data, in a comprehensive and transparent way (Annex, para (a));
- transparent, complete, consistent and accurate information, including methodological information, used at the time of construction of forest reference emission levels and/or forest reference levels including, *inter alia* (as appropriate) a description of datasets, approaches, methods and models, and assumptions used, descriptions of relevant policies and plans, and descriptions of changes from previously submitted information (Annex, para (b));
- pools and gases and REDD+ activities that have been included in forest reference emission levels and/or forest reference levels and the reasons for omitting a pool and/or activity from the construction of forest reference emission levels and/or forest reference levels, noting that significant pools and/or activities should not be excluded (Annex, para (c));
- the definition of forest used in the construction of forest reference emission levels and/or forest reference levels and, if appropriate, in case there is a difference with the definition of forest used in the national greenhouse gas inventory or in reporting to other international organisations – an explanation of why and how the definition used in the construction of forest reference emission levels and/or forest reference levels was chosen (Annex, para (d)).

Technical assessment of submissions from REDD+ countries on their
proposed forest reference emissions levels and/or forest reference levels

COP19 has now provided guidance on how proposed forest reference emissions levels and/or forest reference levels should be assessed.[41] The COP has decided that each submission by a REDD+ country on these proposed levels must be subject to a technical assessment (Decision 13/CP.19, Article 1), the guidelines for which are set out in an Annex. The guidelines are very detailed and relate to the objectives of the technical assessment; its scope; the procedures for the technical assessment; the composition of the assessment team; and the timing for assessment (Annex, Articles 1–18). The COP adopts these guidelines (Article 3) and requests the UNFCCC Secretariat, in accordance with its financial resources, to prepare a synthesis report on the technical assessment process for consideration by the SBSTA after the first year of the assessment (Article 4). Parties to the UNFCCC, particularly developing countries and intergovernmental organisations are invited to nominate experts to the UNFCCC roster of experts (Article 5).

The modalities for MRV

COP19 has decided on the MRV modalities for REDD+ activities, including anthropogenic forest-related emissions by sources and removals by sinks, forest carbon stocks and forest area changes. Essentially, the COP reiterates that this must be done consistently with any existing and future decisions of the COP,[42] while recognising the need to develop the capacities of REDD+ countries in this regard (Decision 14/CP.19, Article 2).

The data and information used by REDD+ countries to estimate their forestry-related emissions must be transparent and consistent over time (Article 3). These data and information should be provided through the biennial update reports required from developing countries under previous COP16 and 17 agreements, although some flexibility is allowed for least developed countries and small island developing States (Article 6). The COP agrees that all emissions reductions from REDD+ activities should be measured in tonnes of CO_{2e} per year (Article 4). The COP encourages countries to improve the data and methodologies used over time, while maintaining consistency with the established, or updated, forest reference emission levels and/or forest reference levels, as set out in COP16 (Article 5).

---

41 Decision 13/CP.19: *Guidelines and procedures for the technical assessment of submissions from Parties on proposed forest reference emissions levels and/or forest reference levels:* http://unfccc.int/files/ meetings/ warsaw_nov_2013/decisions/application/pdf/cop19_frl.pdf.

42 Decision 14/CP.19, Article 1: *Modalities for measuring, reporting and verifying:* http://unfccc.int/ files/ meetings/warsaw_nov_2013/decisions/application/pdf/cop19_mrv.pdf.

Providing a technical annex for expert review

When applying for results-based finance, REDD+ countries are requested to supply a technical annex as set out at COP17, although COP19 now underlines that this is voluntary (Decision 14/CP.19, Article 7).[43] An Annex to this Decision provides the guidelines on preparing a technical annex. These include:

- summary information from the final report containing each assessed forest reference emission level and/or forest reference level which, in turn, includes the levels expressed in tonnes of CO_{2e} per year
- the relevant REDD+ activities being undertaken
- the territorial forest area covered
- the date on which the forest reference emission levels and/or forest reference level was submitted and the date of the final technical assessment report
- the period of years for which the reference levels are assessed (Decision 2/ CP.17, Annex III, Article 1).

The technical annex must also include: the emissions reduction results in tonnes of CO_{2e} per year (Annex III, Article 2); demonstration that the methodologies used to calculate these reductions are consistent with those used to establish the forest reference levels (Annex III, Article 3); a description of the national forest monitoring systems[44] and the institutional roles and responsibilities in-country for MRV;[45] the information that would allow a reconstruction of the results (Decision 14/CP.19, Article 5); and a description of how the requirements set out at COP15 have been taken into account (Article 6).[46]

---

43 The requirements of a technical annex are set out in Decision 2/CP.17 (Annex III): *UNFCCC biennial update reporting guidelines for Parties not included in Annex I to the Convention*: http://unfccc. int/ files/meetings/durban_nov_2011/decisions/application/pdf/cop17_lcaoutcome.pdf.

44 Note that COP19 also established Decision 11/CP.19: *Modalities for national forest monitoring systems*: http://unfccc.int/files/meetings/warsaw_nov_2013/decisions/application/pdf/cop19_ fms.pdf. Here, among others, the COP has decided that national (or subnational) forest monitoring systems should build on existing systems; enable the assessment of different types of forest in the country; be flexible and allow for improvement; and reflect the phased REDD+ approach discussed in Chapter 1 (Decision 2/CP.17, Annex III, Article 4). The systems may also provide information on how the REDD+ safeguards will be addressed and respected (Annex III, Article 5).

45 Decision 11/CP.19, Article 4: *Modalities for national forest monitoring systems*: http://unfccc.int/ files/ meetings/warsaw_nov_2013/decisions/application/pdf/cop19_fms.pdf.

46 Note that these requirements are set out in Decision 4/CP.15, [1(c)] and [1(d)] (UNFCCC 2010). These require REDD+ countries to use the most recent IPCC guidance for estimating anthropogenic forest-related GHG emissions by sources and removals by sinks, forest carbon stocks and forest area changes. They are also required to establish, according to national circumstances and capabilities, robust and transparent national forest monitoring systems and, if appropriate, subnational systems as part of national monitoring systems that:

The COP has also decided that when a developing country Party seeks to receive payments for results-based action, two land use, land use change and forestry (LULUCF) experts on the UNFCCC roster, one each from a developed and developing country, will be included in the technical team of experts that will assess the data and information supplied by a REDD+ country.[47] The COP decided further that the technical team of experts must analyse the extent to which: there is consistency in methodologies, definitions, comprehensiveness and information provided between the assessed reference level and actual results arising from REDD+ activities; the data and information in the technical annex are transparent, complete and accurate, and consistent with the guidelines; and the results are accurate, to the extent possible (Decision 13/CP.19, Article 10).

The COP has also decided that the REDD+ country that has submitted the technical annex may interact with the technical team of experts to provide clarifications and additional information to assist the team's analysis. Also, the two LULUCF experts may seek clarifications, to which the country should respond in accordance with its national circumstances and capabilities (Decision 13/CP.19, Article 12 and 13). The COP has agreed that the LULUCF experts will prepare a technical report to be published by the UNFCCC Secretariat on the UNFCCC REDD+ web platform containing: the REDD+ country's technical annex and the experts' analysis thereof; areas for improvements; and any comments or responses provided by the REDD+ country, including on capacity-building needs (Article 14).

Significantly, the COP has agreed that where any results-based actions are eligible for market-based approaches, such as the generation of REDD+ carbon credits, these may be subject to additional modalities for verification, as the COP deems necessary (Decision 13/CP.19, Article 15).

Results-based financing for REDD+

A *wide variety of funding sources*

COP17 agreed that funding for REDD+ can come from a wide variety of sources – public and private, bilateral and multilateral[48] – and can have

- use a combination of remote sensing and ground-based forest carbon inventory approaches for estimating anthropogenic forest-related GHG emissions by sources and removals by sinks, forest carbon stocks and forest area changes
- provide estimates that are transparent, consistent, as far as possibly accurate and that reduce uncertainties – taking into account national capabilities and capacities
- are transparent and that their results are available and suitable for review as agreed by the COP.

47 Decision 13/CP.19, Article 10: *Guidelines and procedures for the technical assessment of submissions from Parties on proposed forest reference emissions levels and/or forest reference levels*: http://unfccc.int/files/meetings/warsaw_nov_2013/decisions/application/pdf/cop19_frl.pdf.

48 Decision 12/CP.17, Article 65: *Guidance on systems for providing information on how safeguards are addressed and respected and modalities relating to forest reference emission levels and forest reference levels*

a market-based or non-market-based approach (Decision 12/CP.17, Article 66). A market-based approach envisages the generation of carbon credits from REDD+ to be traded on the international market, whereas a non-market-based approach could involve payments by developed to developing countries for the purposes of REDD+.

COP18 decided, with regard to REDD+, to undertake a work programme on results-based finance for REDD+ in 2013 (Draft decision -/CMP.8, UNFCCC 2012, Article 25) to scale up and improve the effectiveness of this finance (UNFCCC 2012, Article 28). The work programme will address the ways and means to transfer payments for results-based action, ways to incentivise non-carbon benefits and ways to improve the coordination of results-based finance (Article 29). The work programme will end by COP19 in 2013 (Article 33).

The 2013 Warsaw Framework for REDD+

Following this, at COP19 under the *2013 Warsaw Framework for REDD+*, a coordinated programme for financial support for REDD+ was established.[49] REDD+ Parties are invited to designate a national entity or focal point to liaise with the UNFCCC Secretariat, and other relevant Convention bodies, on the coordination of financial support for the full implementation of REDD+ activities (Decision 10/CP.19, Article 1). These entities might then be nominated to obtain and receive the results-based payments for REDD+, although these would need to be consistent with any specific operational modalities of the relevant financing entities (Article 2).

The COP recognises that the coordination of support needs to be supplemented by a number of activities by REDD+ countries. These include the sharing of information, knowledge, experience and good practices relating to the implementation of REDD+ at the international level, which would reflect national experiences, and traditional knowledge and practices as appropriate (Decision -/CP.19, Article 3(a)). Any recognised needs and gaps in the coordination of support would have to be identified (Article 3(b)). Information also needs to be exchanged between Convention bodies and other relevant bilateral or multilateral funding bodies, including on the REDD+ obligations placed on countries under COP16, as discussed already (Article 3(c)). It is recognised that REDD+ countries should also provide information, as well as recommendations, to the COP on how the effectiveness of results-based finance, technology and capacity building can be improved to facilitate the implementation of the REDD+ activities – including multilateral, bilateral and private sector finance (Article 3(d) and (e)). Information on the different approaches should

as referred to in decision 1/CP.16, 16: http://unfccc.int/resource/docs/2011/cop17/eng/09a02.pdf#page=16.

49 See Decision 10/CP.19: *Coordination of support for the implementation of activities in relation to mitigation actions in the forest sector by developing countries, including institutional arrangements*: www.redd-monitor.org/wordpress/wp-content/uploads/2013/11/cop19_mitigationactions_forest.pdf.

also be exchanged, including joint mitigation and adaptation approaches for the integral and sustainable management of forests (Article 3(g)).

The REDD+ national entities or focal points are encouraged by the COP to meet on a voluntary basis at the first meeting of SBSTA and the Subsidiary Body for Implementation (SBI) in 2014 to discuss all the elements mentioned earlier regarding the coordination of support (Decision 10/CP.19, Article 4). They are also encouraged to meet formally at the second meeting of these subsidiary bodies and then to meet annually thereafter at the bodies' first sessional meetings (Article 5). The UNFCCC Secretariat is requested to organise these meetings, but only if its budget allows (Articles 6 and 11). COP19 decided that at these meetings, REDD+ countries should seek input from Convention bodies, international and regional organisations, the private sector, indigenous peoples and civil society – including by inviting them to be observers of these meetings (Article 8). The SBI is requested, at its meeting that coincides with the COP in November–December 2017, to review the outcomes of these meetings. This will result in recommendations as to whether any alternatives are required to the existing institutional and governance arrangements for the coordination of support to REDD+ activities (Article 9). It is agreed that the REDD+ work undertaken by the SBSTA and the SBI will conclude at the 2017 COP (Article 10).

The Warsaw work programme on results-based finance

COP19 recalls the previous COP outcomes on REDD+ and recognises the key role that the Green Climate Fund will play in channelling financial resources to developing countries and catalysing climate finance.[50] The COP recalls that REDD+ countries will only receive results-based finance if their REDD+ activities are fully measured, reported and verified, and in accordance with the new COP19 MRV provisions (Decision 10/CP.19, Article 3). Also, before a country can receive results-based payments it must provide its most recent summary of how the REDD+ safeguards have been addressed and respected (Article 4). All sources of funding recognised by the COP are encouraged to channel results-based finance to developing countries in a way that is fair and balanced, while recognising the many different policy approaches that countries might adopt (Article 5). They are also encouraged to apply the methodological guidance provided by the COPs, including COP19, so as to improve the effectiveness and coordination of results-based finance (Article 6). The Green Climate Fund is requested to do likewise (Article 7). Financing entities are also encouraged to continue funding alternative policy approaches such as mitigation and adaptation approaches for the integral and sustainable management of forests (Article 8).

---

50 Decision 9/CP.19, Articles 1 and 2: *Work programme on results-based finance to progress the full implementation of the activities referred to in decision 1/CP, paragraph 70*: http://unfccc.int/files/ meetings/ warsaw_nov_2013/decisions/application/pdf/cop19_redd_finance.pdf.

COP19 has decided to establish an information hub on the web plat-form of the UNFCCC website for the publication of information on countries' REDD+ activities and corresponding results-based finance (Decision 9/CP.19, Article 9), so as to increase the transparency of information on this finance (Article 10). The information to be published includes: the emissions reduction results for each relevant period and the assessed forest reference emission level(s) and/or forest reference level(s) – both to be expressed in tonnes of carbon dioxide equivalent (CO_{2e}) per year – and provided with a link to technical reports now required under COP19; the summary of information on how the safeguards are being addressed and respected; a link to the country's REDD+ national strategy or action plan; and information on the national forest monitoring system as required under COP19 (Article 11). The COP has also decided that the hub must contain information, in consultation with the developing country, on the entity that has paid for results; and the quantity of the emissions reductions paid for, expressed in tonnes of CO_{2e} per year (Articles 12 and 13). The UNFCCC Secretariat is requested, subject to the availability of supplementary resources, to host an expert meeting on the inclusion of information on the hub before the meeting of the SBI at COP20 in December 2014 and to prepare a report on this for the SBI's consideration at COP21 (Article 14). The Secretariat is also requested to improve and further develop the web platform so that the information in REDD+ is easily accessible and available in a simple, transparent manner (Article 19). However, it should only undertake these activities in accordance with the availability of financial resources (Article 24).

The COP notes that the insertion of results on the hub does not create any rights or obligations for any Party or entity and that the information should be linked to any future systems that might be developed under the UNFCCC (Decision 9/CP.19, Articles 16 and 17). The implementation of this decision does also not prejudice any future decisions relating to eligibility or non-eligibility of REDD+ activities (Article 18).

The Standing Committee on Finance is requested as a matter of urgency to focus its earliest possible forum on issues relating to finance for forests – including ways and means to transfer payments for REDD+ activities and the provision of financial resources for alternative approaches (Decision 9/CP.19, Article 20) – and to invite experts to that meeting (Article 21).

Finally, the COP recognises the importance of incentivising non-carbon benefits for the long-term sustainability of implementing REDD+ activities (Article 22).

Ensuring legitimacy

The legitimacy of REDD+ activities is dependent on rigorous technical, environmental, financial and social frameworks. The 'technical' includes the need for forest reference emission levels and/or forest reference levels for each REDD+ country, as well as being able to rely on technology for the purposes of MRV. Without these, there can be no assurance that REDD+ project will deliver the environmental and climate change outcomes it promises. With regard to financial issues, it is still not settled whether public or project-based funding will be regarded as legitimate funding structures. This may vary from country to country. From the social perspective, Article 72 of the AWG-LCA text at COP16, recalled by COP17, attempts to establish the parameters for legitimacy. It reads as follows:

> *Also requests* developing country Parties, when developing and implementing their national strategies or action plans, to address, inter alia, the drivers of deforestation and forest degradation, land tenure issues, forest governance issues, gender considerations and the safeguards identified in paragraph 2 of appendix I to this decision, ensuring the full and effective participation of relevant stakeholders, inter alia, indigenous peoples and local communities.[51]

This exhortation to REDD+ countries is, of course, critically important, for its absence from the REDD+ agreements would have been unthinkable. Nevertheless, it is a somewhat glib statement that fails to reveal the very serious concerns about REDD+ expressed by a range of communities, NGOs and commentators. Developing countries will face considerable challenges 'to address' each of the items raised in Article 72. The drivers of deforestation and forest degradation, as well as forest governance issues, cannot be simply addressed. These aspects have bedevilled the forestry sector for an extensive period of time, while land tenure issues in REDD+ countries are extremely complex. Similarly, the task of ensuring the full and effective participation of all stakeholders, including indigenous peoples and local communities, is rife with difficulties.

Given that this chapter is directed at the legal framework for REDD+, the focus here will be on some of the legal institutions that are necessary for ensuring the legitimacy of REDD+, not including the significant issues associated with land tenure. One crucial question is how the rights of all stakeholders, but particularly indigenous peoples and local communities, will be protected. These rights include the right to participate in decision making around REDD+ schemes, and the right to have their knowledge about forestry resources respected. While the COP16 and 17 REDD+ agreements are 'groundbreaking' for including references to the rights of indigenous peoples and local communities, the rights cannot be simply asserted without undertaking a detailed legal analysis of transparency norms, legal standing and transparent access to decision making in each tropical rainforest country.

---

51 'The Cancun Agreements: 2010 Outcome of the work of the Ad Hoc Working Group on Long-term Cooperative Action under the Convention' (Decision 1/CP.16), Article 72: UNFCCC (2011: 2).

The Rights of Indigenous People

As discussed above, the REDD+ safeguards note the adoption of the UN Declaration on the Rights of Indigenous People.[52] The Declaration states that indigenous peoples shall not be forcibly removed from their lands or territories and may not be relocated without their free, prior and informed consent; and only after agreement on just and fair compensation. Where possible, they must be given the option of return (Article 10). The Declaration requires states to:

- prevent any actions which dispossess indigenous peoples from their land, territories and resources (Article 8(2)(b));
- allow indigenous peoples to participate in decision making on matters which would affect their rights (Article 18);
- consult and cooperate in good faith with the indigenous peoples and to obtain their free, prior and informed consent before adopting and implementing legislative or administrative measures that may affect them (Article 19);
- protect their right to be secure in their means of subsistence and development (Article 20).

Indigenous peoples also have the right to the lands, territories and resources that they have traditionally owned, occupied or otherwise used or acquired (Article 26). Aligned with this are the rights to conservation and protection of the environment and the productive capacity of their lands or territories and resources (Article 29(1)), as well as the right to determine and develop priorities and strategies for their development (Article 32). Many of these rights are replicated and extended in the International Labour Organisation Indigenous and Tribal Peoples Convention (Part II).

Transparency and governance structures: institutional capacity to assert REDD+ rights

Any rights of tenure that indigenous peoples and local communities may have over the carbon sequestered in forests will be ineffectual without the institutional capacity to claim and fully utilise them (Dahal and Adhikari 2008: 20). From a legal and governance perspective, transparency, openness and accountability in government are facilitated by four key pillars: access to information; access to legal remedies before courts and tribunals; the right to participate in government decision making; and access to an independent Ombudsman,[53] or Public Protector,[54] with authority to investigate instances of maladministration in government agencies.

---

52 Available at www.un.org/esa/socdev/unpfii/documents/DRIPS_en.pdf.
53 See, for example, in Australia, the Ombudsman Act 1976 (Cth): www.austlii.edu.au/au/legis/cth/ consol_act/oa1976114/.
54 For information about the South African Public Protector, see http://www.publicprotector.org/.

It must be acknowledged that these institutions are typical of western democracies and may still be emerging in many tropical rainforest countries, although there is no doubt that a strong civil society is recognised as an important contributor to both launching and sustaining a transition from authoritarian to democratic governance (see Antlov et al. 2008). Be that as it may, Global Witness[55] released a full set of Transparency Indicators 2009[56] for making the forest sector transparent.[57] Three of these indicators will be discussed here including: transparency norms, legal standing and transparent access to decision making.

International law and transparency indicators

Principle 10 of the Rio Declaration emphasises the importance of participation, access to judicial proceedings and effective remedies. One of the most comprehensive international law instruments on transparency is the Convention on Access to Information, Public Participation in Decision-Making and Access to Justice in Environmental Matters.[58] The Convention is only binding on countries in Europe; however, it is one of the most comprehensive articulations of what a transparency 'package' encompasses. The Convention: links environmental rights and human rights; establishes that sustainable development can be achieved only through the involvement of all stakeholders; links government accountability and environmental protection; and focuses on interactions between public authorities and the public in a democratic context. Specifically, it requires all Parties to 'endeavour to ensure that officials and authorities assist and provide guidance to the public in seeking access to information, in facilitating participation in decision-making and in seeking access to justice in environmental matters' (Article 3(2)).

Transparency norms

According to Global Witness, with regard to transparency norms, the following questions need to be asked:

* Is there a Freedom of Information (FOI) Act?
* What other rules provide for transparency?
* Are there any sector specific laws, rules or statements that provide for transparency?
* Is there any settlement process for disputes regarding access to information?

---

55 See www.globalwitness.org/.
56 See www.foresttransparency.info/background/2009/methodology/212/full-set-of-transparency-indicators-2009/.
57 See www.foresttransparency.info/cms/file/242.
58 See www.unece.org/env/pp/documents/cep43e.pdf.

The value of FOI legislation is that those impacted by government decision making can be fully informed about the government's actions. Without this information it is difficult for people to exercise their rights and make informed choices. With respect to REDD+, it is essential for indigenous peoples and local communities to be able to access information about: where REDD+ sites will be established; who will manage the sites; how they will be impacted by the establishment of sites and the legal obligations they will have within those sites; what financial benefits will be distributed for managing REDD+ sites; and, importantly, what financial benefits they are likely to receive. In most instances, FOI legislation requests for information are made in writing to government agencies and a dedicated officer is appointed within the agency to comply with requests. FOI legislation often contains exemptions that allow agencies to refuse requests such as 'commercial in confidence' or if it would affect 'law enforcement and public safety'.[59] Where requests for information are refused, an internal appeal process is often established within the government agency itself, with further rights of appeal being available to an independent tribunal, which can overrule the agency's refusal to grant the request for information.[60]

There is evidence that FOI legislation is being adopted in new democracies. This emerges with the enactment in Indonesia of a Freedom of Information Law in 2008, which came into effect on 30 April 2010, and which is discussed in Chapter 8.

Although Global Witness does not specifically mention an Ombudsman or Public Protector, this is an additional transparency institution that might be considered in the context of REDD+. Generally, such an office provides one of the most inexpensive and informal accountability mechanisms to people aggrieved by maladministration in government. An affected individual simply contacts the office to lodge a complaint. The Ombudsman then conducts an investigation, has open access to government agency offices, documents and records and attempts to resolve the dispute between the individual and the agency. The Ombudsman might make recommendations to the agency. The Ombudsman's final sanction is the lodging of a report in parliament, which exposes the instance of maladministration in the government agency. The flexibility and informality of procedures offered by such an office has obvious benefits in the context of REDD+. This is discussed in the Indonesian context in Chapter 8.

Legal standing

Legal standing is the term used to describe permission, granted either by the courts or statute, to appear before the courts in order to litigate. In the context

---

59 See, for example, Schedule 1 of the Freedom of Information Act 1989 (NSW): www.austlii.edu.au/cgi-bin/disp.pl/au/legis/nsw/num_act/foia1989n5249.pdf.

60 *Ibid.*, s 53.

of REDD+ rights against government, it is the right to bring an action against the government challenging the legality of its decision making. This is known as an application for judicial review, and in most cases will test whether a government decision maker has acted consistently with legislation enacted by the parliament – such as forestry legislation. In some jurisdictions, governments provide an additional right to citizens by establishing administrative tribunals. These tribunals do not test the legality of a decision, but rather the correctness. In other words, applicants will argue that the government's decision, say, to grant a logging permit, or to establish a REDD+ site, has been wrongly made. The tribunal then has the authority to 'step into the shoes' of the government decision maker, reassess all the facts, call for new evidence and either confirm the decision or substitute it with its own preferred decision.

Many arguments have been advanced for why courts have the right to determine who can bring a legal action, such as not wanting to open 'the floodgates' of litigation or not wanting to give a forum to 'vexatious litigants' and 'busy bodies'. In other words, courts have wanted to ensure that litigants have a direct personal interest in the subject matter of the litigation. In the context of REDD+, it would be necessary to investigate rights of standing in each tropical rainforest country in order to assess the standing of indigenous peoples and local communities to assert their rights. It seems likely that they would have no difficulty establishing that they are directly impacted by government action in forests.

More problematic is the question of whether public interest actions are permitted. These arise when a public interest law firm, an NGO or an individual seeks to bring an action on behalf of those unable to represent themselves; or simply on behalf of their own members, which seek to protect the world's forests; or because the individual seeks to right a public wrong. Ultimately, it is a matter for the courts and parliament in each tropical rainforest jurisdiction to determine what the standing provisions are. However, as this discussion has demonstrated, a lack of standing can mean that challenges to government decisionmaking around REDD+ cannot be advanced before the courts. Legal standing in Indonesia is discussed in Chapter 8.

The need for full and effective participation of indigenous peoples and local communities: transparent access to decision making

The right, and perhaps even the duty, to participate in government decision making is well established in the literature (see, for example, Sunstein 1988). As discussed earlier, the REDD+ safeguards require respect for the knowledge and rights of indigenous peoples and members of local communities; and the need for the full and effective participation of relevant stakeholders – including, in particular, indigenous peoples and local communities. With regard to indigenous peoples, the principle of free, prior and informed consent (FPIC) is key, as demonstrated by its inclusion in more than one Article of the Declaration. In practice, however, application of the FPIC principle is reportedly very limited among current REDD+ projects in Indonesia (Morgan 2010). A statement, posted on the REDD Monitor website (Lang 2011) and signed by the Central Kalimantan Chapter of the Alliance of Indonesian Indigenous Communities (AMAN, an active indigenous rights advocacy NGO) argues that the REDD pilot projects in Central Kalimantan such as the AU\$30 million AusAID-Indonesia Kalimantan Forests and Climate Partnership (KFCP) have ignored the FPIC principle, are not transparent and are operating in a legal/policy 'grey area'. This is the result of the lack of a provincial REDD+ strategy and lack of clear legal division of authority between heads of government and indigenous institutions. The AMAN statement also notes that in Central Kalimantan, local by-laws only regulate individual land rights, not collective indigenous rights to land.

With regard to effective participation, Global Witness states that some of the key questions to be asked are:

- Is there a national forest forum?
- Are there local forest forums?
- Is there a procedure for consultation on new norms?
- Is there an established list of stakeholders?
- Are reports on consultation processes public?

Although these are key questions, they are linked to forest tenure reform in landscapes, which are often composed of 'multiple stakeholders, competing interest groups and distinct public agencies holding rights and claiming control over land and forest resources. The transfer of rights is further complicated because it is multifaceted, involving different types of tenure systems and rights bundles' (Cronkleton et al. 2010). Under forest tenure reform, rights are often transferred to community-level stakeholders who provide detailed local knowledge essential to good forestry management. Co-management rights are often granted, giving rights holders decision-making power. Essentially, legislation passed to facilitate participation needs to identify: 'who participates in management and how they participate; which resources will be managed and how; and who benefits from management and how' (Cronkleton et al. 2010: 46–7).

All of this ought to be set out in specific forestry legislation, which moves 'transparent access to decision making' from an abstract concept to an established right on the part of indigenous peoples and local communities.

It is often the case that although legislation seems to enshrine the right to participate, in practice, these legislative imperatives are simply ignored by governments or by donors.[61] Clearly then, effective rights of participation

61 This has been a concern in Indonesia, for example, where the following legislation provides quite extensive participation requirements: Law 26 on 2007 on Spatial Planning; Law 32 of 2009 on Environmental Management and Protection; National Environment Minister Decision 24 of 2009 on the Guidelines for Evaluating Environmental Impact Assessment Documents; Head of the National Body for Environmental Impact Control Decision 8 of 2000.

in REDD+ decision making will depend in large measure on the political space and freedom that civil society enjoys in any given jurisdiction. There is evidence that international donors have targeted civil society strengthening as integral to realigning State–society relations, so as to expand citizen participation and reinforce State responsiveness and accountability (see Antlov et al. 2008). Evidence in Indonesia, for example, is that over the past decade, the number of civil society organisations have grown dramatically as democratisation has created the space for them to establish rights and mechanisms of accountability. Regional autonomy and decentralisation have enhanced these opportunities (Antlov et al. 2008: 4). Yet prominent civil society leaders still lament the fact that 'the guardians of justice are corrupt' so that the transition to democratic consolidation is incomplete (Antlov et al. 2008: 2). Public participation and corruption in Indonesia are discussed in Chapter 4.

Conclusion

There is no doubt that the decisions from the Conferences of the Parties to the UNFCCC and the Meeting of Parties to the Kyoto Protocol since Bali have attempted to deal with the concerns that have been raised about the legitimacy of REDD+ projects. This chapter has indicated where the concerns lie and, from a legal perspective, how they might be addressed. However, given that the international law mechanisms for REDD+ remain at an incipient stage, it will take time and a great deal of research to discover whether all of the safeguards and other legitimacy criteria are effectively implemented by the REDD+ countries. It is with a view to the concerns raised in this chapter that we now turn now to investigate the prospects for implementation of the REDD+ safeguards in the Indonesian context.

References

Antlov, H., D. W. Brinkerhoff and E. G. Rapp (2008) *Civil society organizations and democratic reform: Progress, capacities and challenges in Indonesia*, RTI International: www.rti.org/pubs/Antlov_CSOs_in_Indo_ARNOVA.pdf.
Cronkleton, P., D. Barry, J. M. Pulhin and S. Saigal (2010) 'The devolution of management rights and the co-management of community forests', in A. Larson, D. Barry, G. R. Dahal and C. J. P. Colfer (eds) *Forests for people: Community rights and forest tenure reform*, London: Earthscan.
Dahal G. and K. Adhikari (2008) 'Trends and impact of forest tenure reforms in Asia: Cases from India, Indonesia, Lao PDR, Nepal and the Philippines', *Journal of Forests and Livelihood* 7(1): 20.
Dooley, K. (FERN) and L. Horner (Friends of the Earth US) (2012) *FW special report – Durban aimed to save the market, not the climate, December 2011*, EU Forest Watch, January: www.redd-monitor.org/wordpress/wp-content/uploads/2012/01/Durban-update.pdf.
Fisher, R. and R. Lyster (2013) 'Land and resource tenure: The rights of indigenous peoples and forest dwellers', in R. Lyster, C. MacKenzie and C. McDermott C.

(eds) *Law, tropical forests and carbon: The case of REDD+*, New York: Cambridge University Press.

Lang, C. (2011) 'Indigenous peoples' organisation demands "immediate moratorium" on REDD+ in Central Kalimantan', *REDD-Monitor*, 22 June: www.redd-monitor. org/ 2011/06/22/indigenous-peoples-organisation-demands-immediate-moratori um-on-redd-in-central-kalimantan/#more-8963.

Lyster, R. (2009) 'The new frontier of climate law: Reducing Emissions from Deforestation (and Degradation)', *Environmental and Planning Law Journal* 26(6): 417: http://papers.ssrn.com/sol3/papers.cfm?abstract_id=1531990.

Lyster, R. (2011) 'REDD+, transparency, participation and resource rights: The role of law', *Environmental Science & Policy* 14(2): 118.

Maulia, E. and H. D. Tampubolon (2010) 'Government not ready for information law rollout', *Jakarta Post*, 30 April.

Morgan, B. (2010) 'REDD at the community level: Community engagement and carbon conservation in Indonesia's forests', unpublished Master's thesis.

Streck C. and S. M. Scholz (2006) 'The role of forests in global climate change: Whence we come and where we go', *International Affairs* 82(5): 868.

Subsidiary Body for Scientific and Technological Advice (SBSTA) (2007) *Reducing emissions from deforestation in developing countries: Draft conclusions proposed by the Chair*, Twenty-sixth session, Bonn, Germany: 7–18 May 2007, UN Doc. No. FCCC/ SBSTA/2007/L.10, published 17 May: www.rainforestcoalition.org/documents/ UNFCCCSBSTA2007l10.pdf.

Subsidiary Body for Scientific and Technological Advice (SBSTA) (2011) *Methodological guidance for activities relating to reducing emissions from deforestation and forest degrad-ation and the role of conservation, sustainable management of forests and enhancement of forest carbon stocks in developing countries*, Thirty-fifth session, Durban, South Africa: 28 November to 3 December 2011, UN Doc. FCCC/SBSTA/ 2011/L.25/Add.1: http://unfccc.int/resource/docs/2011/sbsta/eng/l25a01.pdf.

Sunstein, C. (1988) 'Beyond the Republican revival', *Yale Law Review* 97(8): 1538.

United Nations Framework Convention on Climate Change (UNFCCC) (2002) *Report of the Conference of the Parties on its seventh session, held at Marrakesh from 29 October to 10 November 2001*, UN Doc. FCCC/CP/2001/13/Add.1, published 21 January: http://unfccc.int/resource/docs/cop7/13a01.pdf#page=54.

United Nations Framework Convention on Climate Change (UNFCCC) (2009) *Draft decision -/CP.15: Proposal by the President – Copenhagen Accord*, Conference of the Parties, Fifteenth session, Copenhagen, Denmark, 7–18 December 2009, UN Doc. FCCC/CP/2009/L.7, published 18 December: http://unfccc.int/resource/ docs/2009/ cop15/eng/l07.pdf.

United Nations Framework Convention on Climate Change (UNFCCC) (2010) *Report of the Conference of the Parties on its fifteenth session, held in Copenhagen from 7 to 19 December 2009*, UN Doc. FCCC/CP/2009/11/Add.1, published 30 March 2010: http://unfccc.int/resource/docs/2009/cop15/eng/11a01.pdf.

United Nations Framework Convention on Climate Change (UNFCCC) (2011) *Report of the Conference of the Parties on its sixteenth session, held in Cancun from 29 November to 10 December 2010*, UN Doc. FCCC/KP/CMP/2010/12/Add.1, published 15 March: http://unfccc.int/resource/docs/2010/cop16/eng/07a01.pdf.

United Nations Framework Convention on Climate Change (UNFCCC) (2012) *Outcome of the work of the Ad Hoc Working Group on Further Commitments for Annex I Parties under the Kyoto Protocol*, Conference of the Parties serving as the meeting of the Parties to the Kyoto Protocol, Eighth session, Doha: Qatar, 26 November to 7 December 2012, UN Doc. FCCC/KP/CMP/2012/L.9, published 8 December: http://unfccc.int/documentation/documents/advanced_search/items/6911.php?priref=600007290.

3 Indonesia and international climate law and policy

This chapter examines Indonesia's engagement with the development of the international climate regime, with particular focus on Indonesia's attitude in international forums towards the evolution of the REDD+ mechanism.

The future of the international climate regime remains uncertain as States continue to work towards the objective of the Durban Platform for Enhanced Action, agreed at United Nations Framework Convention on Climate Change (UNFCCC) seventeenth session of the Conference of the Parties (COP17) in Durban in December 2012, 'to develop a protocol, another legal instrument or an agreed outcome with legal force under the Convention applicable to all parties'[1] to enter into force from 2020. In lieu of agreement on a global climate regime embracing all major emitters, there have been important achievements in specific areas of carbon management. One of the most notable of these is in relation to forests, as seen in the adoption and evolution of REDD+ initiatives.

As noted in Chapter 2, proposals for REDD+ originated on the international level in a proposal by Papua New Guinea and Chile at COP11, held in Montreal in 2005. COP11 marked the entry into force of the Kyoto Protocol and the first meeting of the Parties to the Kyoto Protocol. There has been a strong Indonesian connection with the subsequent development of REDD+. It is significant that it was at COP13, held in Bali in December 2007, that agreement was reached in the Bali Action Plan to explore '[p]olicy approaches and positive incentives on issues relating to reducing emissions from deforestation and forest degradation in developing countries; and the role of conservation, sustainable management of forests and enhancement of forest carbon stocks in developing countries'.[2] REDD+ was also referred to in Decision 2/CP.13 on *Reducing Emissions from Deforestation in Developing Countries: Approaches to stimulate action*, which set out a framework for the emergence of REDD+ in subsequent COPs and requested the SBSTA to begin a programme of work on the methodological issues connected with REDD+ (FIELD 2012).

---

1 See http://unfccc.int/files/meetings/durban_nov_2011/decisions/application/pdf/cop17_durban platform. pdf.
2 Bali Action Plan, [1(b)(iii)].

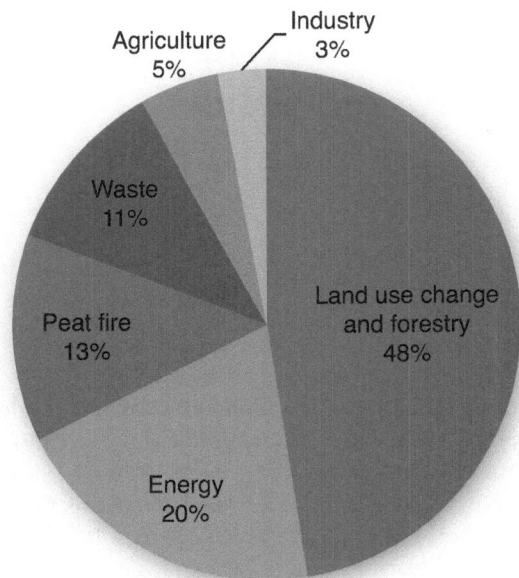

Figure 3.1 Indonesia's national emissions contributions by sector, 2000
Source: Government of Indonesia 2011, *Second National Communication to the UNFCCC*: II-7

Indonesia's greenhouse gas emissions profile

When all sectors are included, Indonesia is one of the largest greenhouse gas emitters (the fifth largest, after China, the United States, the European Union (EU) and Brazil (World Resources Institute 2013)). The Indonesian government has been acutely aware of this and former President Yudhoyono demonstrated a strong commitment to reducing emissions as one component of Indonesia's 'green growth agenda'. Around two-thirds of Indonesia's emissions are produced by land use change through deforestation, draining and burning of peatlands and agriculture (Burke and Resosudarmo 2012: 311). Emissions from energy production are a relatively small component of Indonesia's total emissions profile, but are increasingly rapidly (an average of 4.4 per cent per annum) as a result of sustained economic growth over the last decade (Burke and Resosudarmo 2012: 311–12) (see Figure 3.1).

Indonesia and the UNFCCC and Kyoto Protocol[3]

In 1988 the United Nations General Assembly recognised climate change as a 'common concern of mankind' and initiated a process of scientific analysis and international attention that led ultimately to the conclusion of the UNFCCC and Kyoto Protocol. The overarching goal of the climate change regime is

3 Parts of this discussion are drawn from Saul et al. (2012: Ch. 3).

set out in Article 2 of the UNFCCC: the 'stabilization of greenhouse gas concentrations in the atmosphere at a level that would prevent dangerous anthropogenic interference with the climate system.' The provision goes on to stipulate that this level should be achieved within a time frame that will permit ecosystems to adapt naturally, prevent threats to food production and to ensure that economic development can proceed sustainably.

The UNFCCC divides countries into two categories: industrialised nations, including economies in transition such as eastern European States; and non-industrialised nations, which are the developing States that make up the majority of the international community. The industrialised group of countries are listed in Annex I of the UNFCCC and so are usually known as 'Annex I countries'. The UNFCCC imposes some obligations on all parties, such as duties to monitor and report their greenhouse gas emissions and develop programmes to mitigate climate change. But it is the Annex I states that are given special duties on the 'basis of equity' and in accordance with principle of 'common but differentiated responsibilities and capabilities' to 'take the lead in combating climate change and the adverse effects thereof'.

The Kyoto Protocol operationalises the central bargain agreed in the UNFCCC, namely that developed countries would take on legally binding emissions reductions or limitation targets and provide technological and other assistance to developing countries to help them develop less carbon-intensive economies. While undoubtedly a major achievement in global environmental governance, in truth the Kyoto Protocol was a very modest first step towards combating climate change. When Kyoto was negotiated it was anticipated that the first commitment period would be followed by second and subsequent periods and with Australia (Combet 2012) and several other States[4] joining a second commitment period to Kyoto, there is an opportunity for agreement to be reached on a new round of emissions reductions spanning the period 2013 to 2020.

Indonesia has been strongly supportive of the international climate regime since its inception in 1992. One key reason is that Indonesia is located in the tropical zone, thus is highly vulnerable to the effects of climate change, with increasing temperatures affecting the productivity of agriculture, damaging reefs and fisheries. The sea level rise associated with climate change threatens many parts of the Indonesian archipelago, including the major cities of Jakarta and Surabaya through impacts such as saltwater intrusion and the loss of agricultural areas (Garnaut 2009: 109).[5]

---

4 This includes Belarus, Croatia, members of the EU, Iceland, Kazakhstan, Liechtenstein, Monaco, Norway, Switzerland and Ukraine. Several major emitters have indicated they have no intention of joining a second Kyoto commitment period, including Japan, the Russian Federation and Canada (which withdrew from the Kyoto Protocol in December 2011). See UNFCCC (2012a).

5 However, Indonesia's climate change vulnerability is lower than other comparable countries, including other populous Muslim countries, which, except for Indonesia, are all found in high-vulnerability regions in Africa, the Middle East and South Asia (Diamond-Smith 2011).

Indonesia signed the UNFCCC in 1992 and ratified the Convention on 23 August 1994, pursuant to Law 6 of 1994 on the Ratification of the UNFCCC. Indonesia has also joined other key global treaties in relation to atmospheric pollution, including the 1985 Vienna Convention for the Protection of the Ozone Layer and the 1987 Montreal Protocol on Substances that Deplete the Ozone Layer.[6]

Indonesia's attitude to the UNFCCC in the initial phase of the international climate regime's evolution was set out in its 1997 Report to the United Nations Commission on Sustainable Development:[7]

> Nations endorsing the UN Framework Convention on Climate Change (UNFCCC) in Rio de Janeiro in June 1992 recognized that climate change is a potentially major threat to the world's environment and economic development. As one of the 154 States to the Framework Convention on Climate Change, Indonesia is well aware of the issue.
>
> Indonesia is also acutely aware that global temperature change might result in sea level rise. Such a rise in sea level will have serious consequences for Indonesia as an archipelagic country with 17,500 islands and a coastline of more than 81,000 kilometers. The industries, infrastructure, urban populations and most fertile agricultural lands are concentrated in low lying coastal areas. Of a total of about 180 million Indonesians, approximately 110 million live in coastal areas. Indonesia will therefore suffer significant physical and socio-economic impacts from even very small rises in sea level ...
>
> Indonesia has recognized the importance of climate change as an international, regional and national issue. As a consequence, Indonesia has participated in the work of the Intergovernmental Panel on Climate Change (IPCC) and the Intergovernmental Negotiation Committee (INC) for the UN Framework Convention on Climatic Change since they were established. To build a strong basis for further responses on climate change, Indonesia has signed the United Nations Framework Convention on Climate Change along with 153 other countries. This Convention has been ratified through Act Number 6 (1994) of the Republic of Indonesia.
>
> The ratification of the Convention will act as an instrument of change within Indonesia and abroad. Within the country, this convention will add legal instruments to help implement environmentally sustainable development in relation to climatic issues. Abroad, it shows that Indonesia has taken an active role in global environmental issues, and it

---

6   Indonesia acceded to the Vienna Convention and ratified the Montreal Protocol on 26 June 1992 and has acceded to, or ratified, all subsequent amendments to the ozone regime (the London Amendment, Copenhagen Amendment, Montreal Amendment and Beijing Amendment): http://ozone.unep.org/ new_site/en/treaty_ratification_status.php.

7   See http://www.un.org/esa/earthsummit/indon-cp.htm.

gives more and greater opportunities for Indonesia to work together and to communicate with other countries and organizations.

Indonesia has also signed and ratified the Kyoto Protocol. It signed the Protocol on 13 July 1998, but it was not until 3 December 2004 that it ratified the treaty. The Kyoto Protocol entered into force generally, including for Indonesia, on 3 March 2005. Several reasons have been given for the delay between Indonesia's signature and ratification of the Kyoto Protocol, the most crucial being the decision by the United States, in March 2001, not to proceed to ratification. As Murdiyarso expresses, the Indonesian position is '[i]f such a wealthy nation with the largest emissions decided to oppose the Protocol, why should we bother?' (Murdiyarso 2004: 145–6) Indonesia ultimately did join the Kyoto Protocol, following an extensive national process of implementing the UNFCCC and preparing for the Kyoto Protocol through 'capacity-building activities by hosting various projects, including the Activities Implemented Jointly pilot phase, a national action plan, institution strengthening, assessment of technologies, national strategy studies, and the completion of the First National Communication' (Murdiyarso 2004: 146) with the support of funding from individual States such as Australia and institutions such as the World Bank.

Indonesia has fulfilled its reporting obligations as a Non-Annex I party to the UNFCCC, submitting its first national communication in October 1999 and the second national communication in January 2011 (which was subsequently updated in 2012 (Government of Indonesia 2011)). It was Indonesia's 'focused and effective hosting of the United Nations climate change conference in Bali in December 2007' (Garnaut 2009: 112) that served to highlight to the international community the Indonesian government's commitment to controlling global warming. Indonesia has played an active role in subsequent COPs at Poznan, Copenhagen, Cancun, Durban, Doha and Warsaw.

In 2007 Indonesia adopted a 'National Action Plan for Greenhouse Gas Emission Reduction' and in 2010 Indonesia submitted its Nationally Appropriate Mitigation Actions (NAMAs) when it formally confirmed its association with the Copenhagen Accord.[8] The Copenhagen conference came at the end of two years of negotiations that began in 2007 and concluded with the Bali Action Plan. At previous climate conferences, the main division has been between developed nations and developing nations – between the north and the south. In Copenhagen, there was a new fracture, between a small group of developing and least developing States and larger developing States such as China and India with rapidly growing emissions. The divisions proved insurmountable and Copenhagen led to no legally binding outcome whatsoever. Instead, what was agreed was merely a 12-paragraph political 'accord' – the Copenhagen Accord – negotiated directly between President Barack

8 See http://unfccc.int/files/meetings/cop_15/copenhagen_accord/application/pdf/indonesiacphac-cord_ app2.pdf.

Obama, Chinese Premier Wen Jiabao, Indian Prime Minister Manmohan Singh, Brazilian President Luiz Inacio Lula and South African President Jacob Zuma. The Accord was not formally adopted by the COP to the UNFCCC or the Meeting of the Parties (MOP) to the Kyoto Protocol. Instead, the COP and the MOP adopted decisions that they 'take note' of the Accord, leaving governments then to sign on to the document if they wished to do so.

In its NAMAs communicated to the UNFCCC as part of the Copenhagen process, Indonesia committed to an emissions reduction of 26 per cent by 2020 from business as usual, through the following actions:

1   Sustainable peatland management.
2   Reduction in rate of deforestation and land degradation.
3   Development of carbon sequestration projects in forestry and agriculture.
4   Promotion of energy efficiency.
5   Development of alternative and renewable energy sources.
6   Reduction in solid and liquid waste.
7   Shifting to low-emission transportation mode.

At COP17 in Durban, December 2011, the 'Durban Platform for Enhanced Action' was agreed. While there was no agreement on emissions reductions for a second Kyoto commitment period, the Durban Platform was nonetheless important because it called for developed and developing countries to reduce their emissions and set a timeline for a new round of negotiations to be concluded by 2015 on a 'protocol, another legal instrument or an agreed outcome with legal force' that would have effect from 2020.

At COP18 in Doha, December 2012, Rachmat Witoelar, President Yudhoyono's Special Envoy for Climate Change and Head of the Indonesian Delegation, set out Indonesia's position on the future of the climate change regime being negotiated under the Durban Platform (see UNFCCC 2012b).[9] He acknowledged progress made at Cancun and Durban, but noted that there was still much 'unfinished business' and that 'Indonesia is of the view that developed countries should show their leadership in reducing emissions.' Mr Witoeler went on to highlight three elements that Indonesia considered important for Doha to deliver:

1   The adoption of a second commitment period for the Kyoto Protocol.
2   The termination of the mandate of the Ad Hoc Working Group on Long-Term Cooperative Action under the UNFCCC (the AWG-LCA).
3   That the Ad Hoc Working Group on the Durban Platform for Enhanced Action (ADP) should continue working on a plan to address the 'pre-2020 ambition gap and [deliver] a new legally binding agreement by 2015'. It was emphasised that 'the work under the ADP must not lead

---

9 See   http://unfccc.int/resource/docs/cop18_cmp8_hl_statements/Statement%20H.E%20Mr. Rachmat%20Witoelar,%20Indonesia.pdf

to the reinterpretation of the key principles of the Convention, especially the principles of equity, common but differentiated responsibility, and respective capabilities'.

These three objectives were largely achieved at Doha, at which Indonesia played a fairly low-key role. In setting out these three objectives, Indonesia was speaking on its own behalf and for the Cartagena Dialogue for Progressive Action (Cartagena Dialogue). Indonesia is a member of two, partially overlapping, country groupings within the UNFCCC:

- the Coalition for Rainforest Nations
- the Cartagena Dialogue, which is a group of developing and developed UNFCCC members established in a meeting in Cartagena, Colombia, following COP15 in Copenhagen in 2009.[10]

At COP18, agreement was reached on a second commitment period for the Kyoto Protocol, bringing to a culmination and completion the work of the Ad Hoc Working Group on Further Commitments for Annex I Parties under the Kyoto Protocol (AWG-KP) which was established at MOP in Montreal in 2005. Thirty-five States, including Australia, agreed to emissions reduction targets under a second commitment period for the Kyoto Protocol (UNFCCC 2012b). The second commitment period will run from 1 January 2013 until 31 December 2020, three years longer than the first commitment period (2007–12).

The other negotiating track under the climate regime, the AWG-LCA, was also brought to a conclusion at COP18. The AWG-LCA was established as a subsidiary body at COP13 in Bali, essentially to bring key States outside the Kyoto Protocol, most notably the United States, into a multilateral process to agree on 'long-term cooperative action' to address climate change. The AWG-LCA was conceived as a core component of the 'Bali Road Map' – a two-year process that sought to achieve a new climate regime at COP15 in Copenhagen. However, as noted above, that did not occur at Copenhagen and the parties agreed to extend the mandate of the AWG-LCA for another year. At COP16 and again at COP17 the AWG-LCA was extended, with a final end date given as COP18 in Doha. The main achievements of the AWG-LCA have been the in-principle agreement that both developed *and* developing States must take on mitigation efforts.

The AWG-LCA became effectively redundant following agreement at COP17 in Durban that work would begin on a comprehensive climate treaty, to be concluded by 2015. The negotiations on the new regime are taking place within the ADP. Among developed States, it is anticipated that the ADP will help usher in 'an evolving and dynamic framework that reflects current

---

10 See www.climatechange.gov.au/about/accountability/annual-reports/annual-report-1011/feature-cartagena.aspx.

socio-economic realities and definitively dismantles the "firewall" between developed and developing country mitigation' (Allan et al. 2012: 28).

At COP19 in Warsaw in December 2013, Indonesia expressed its support for the outcomes of the conference, which included a work plan for the 2015 Agreement, the Warsaw Framework for REDD+, the Warsaw International Mechanism for Loss and Damage, decisions on a funding mechanism under the UNFCCC and general guidelines on monitoring, reporting and verification (MRV) to support mitigation action in developing countries or NAMAs. However, Indonesia's lead negotiator, Rachmat Witoelar, in an unusual and pointed remark, registered his strong concern that the new government in Australia was abolishing all of its 'governmental policies on climate change' (Government of Indonesia 2013).

Indonesia and the Clean Development Mechanism

Indonesia enjoys significant opportunities to generate national income from the reduction of greenhouse emissions and the sale of carbon credits generated as a result on international markets. Indonesia's primary focus for generating carbon credits has been through REDD+ schemes rather than the Clean Development Mechanism (CDM). In 2012 Indonesia had 80 CDM projects on foot, a relatively small number compared with other similar developing States such as Malaysia and Vietnam (Burke and Resosudarmo 2012: 314). Indonesia's CDM projects are in the following sectors: biogas, methane avoidance, biomass, methane recovery, hydroelectric power, cement, renewable energies, fuel switch, energy efficiency, N_2O decomposition, PFC reduction and substitution as well as waste gas or heat utilisation (Takahashi and Kuriyama 2013).

Indonesia and REDD+

It is well known that the Kyoto Protocol's CDM mechanism does not permit a wide range of forestry activities for generating CDM carbon credits (known as certified emission reduction units (CERs)). In the forestry sector only limited afforestation and reforestation projects are permitted CDM activities. Indonesia has not sought to make use of this aspect of the CDM mechanism as it is deforestation that remains its main forestry policy challenge. Accordingly, Indonesia's support for REDD+ and other forestry and land use-related approaches to mitigating climate change should be viewed against the backdrop of Indonesia's emissions profile and the opportunity for major reductions at relatively low cost. Indonesia holds the world's second largest stocks of forest carbon (behind Brazil (Saatchi et al. 2011: 9902). As Garnaut (2009: 112) has explained:

> The global community and Indonesia both have strong interests in introducing incentives for greenhouse gas abatement to take place at low cost

in Indonesia rather than at higher cost elsewhere. The opportunities for low-cost abatement cover afforestation and re-afforestation as well as avoided deforestation. Working with developed countries to introduce these incentives could be a special Indonesian contribution to the global mitigation effort.

In its Second National Communication to the UNFCCC, the Government of Indonesia (2011: V-20) set out the five key national policies in the forestry sector:

1   Combating illegal logging and its associated trade.
2   Revitalisation of the forestry sector.
3   Conservation and rehabilitation of forest resources.
4   Empowering the economy of the community within and surrounding the forest area.
5   Stabilisation of forest area for promoting and strengthening sustainable forest management.

Deforestation has been identified as Indonesia's 'most pressing environmental concern' (Burke and Resosudarmo 2012: 314) given the rapid rate at which its forests are being lost as a result of pressures for agricultural land, palm oil plantations and the timber trade. As much as 0.5 per cent of Indonesia's total forests were lost in 2011 alone, placing Indonesia above Brazil and the Democratic Republic of the Congo in terms of the rate of deforestation (Burke and Resosudarmo 2012: 314).

Indonesia and complementary climate change measures

In addition to being actively engaged in the international climate regime, in particular in development of the REDD+ mechanism, Indonesia has also supported complementary measures on a bilateral and multilateral basis.

In 2007 Indonesia began negotiations with the EU as part of the EU's Forest Law Enforcement, Governance and Trade (FLEGT) Action Plan to address illegal logging in the world's forests. Under the FLEGT nations may enter into voluntary partnership agreements (VPAs) with the EU. These define 'legal timber', assist in the analysis of existing legislation and, where necessary, identify possible legal reforms in timber-producing countries (Brack 2012: 211). Indonesia has expressed a strong preference for voluntary, mutually supportive arrangements of this type to address illegal logging, rather than unilateral measures such as Australia's recently enacted Illegal Logging Prohibition Act 2012 (Cth) (Saul and Stephens 2012). In 2011 Indonesia and the EU signed a VPA, which once ratified and implemented will be a legally binding agreement to improve forest sector governance in

Indonesia and ensure that timber products exported from Indonesia to the EU are produced in compliance with Indonesian law.

Future opportunities for Indonesia in the international climate regime

Although Indonesia has been a supporter of the international climate regime and a particularly active player in a specific sector of global mitigation efforts (forestry and land use change), it has not been at the forefront of proposals to develop and expand on the climate change regime. There is an opportunity for Indonesia, working with developed country partners such as its nearest neighbour, Australia, to assume a more prominent role in articulating a clear vision for further development of the climate regime. As discussed above, Australia and Indonesia have cooperated closely in the forestry context, including through joint submissions on REDD+ issues to climate negotiations. Garnaut has argued that this cooperation could go further and that Indonesia and Australia could conceivably develop 'a set of principles for allocating the global mitigation effort across countries' that would be seen as 'fair and practical' (Garnaut 2009: 113).[11] Indonesia is well placed to achieve this given its prominence within the Cartagena Dialogue, that brings a group of developing and developed countries together that have a shared interest in achieving a comprehensive and legally binding climate regime that crosses the north–south divide.

References

Allan, J., B. Antonich, A. Appleton, R. R. Bhandary, K. Kulovesi, R. Kosolapova et al. (2012) 'Summary of the Doha climate change conference', *Earth Negotiations Bulletin* 12: 1.

Brack, D. (2012) 'Excluding illegal timber and improving forest governance: The European Union's forest law enforcement, governance and trade initiative' in P. Lujala and S. A. Rustad (eds) *High-value natural resources and post-conflict peace-building*, London: Earthscan.

Burke, P. J. and B. P. Resosudarmo (2012) 'Survey of recent developments', *Bulletin of Indonesian Economic Studies* 48: 299.

Combet, G. (2012) 'Australia ready to join Kyoto second commitment period', media release, 9 November.

Diamond-Smith, N. (2011) 'Climate change and population in the Muslim world', *International Journal of Environmental Studies* 68: 1.

Elliott, L. (2011) 'Australia, climate change and the global south', *The Round Table* 100: 441.

Foundation for International Environmental Law and Development (FIELD) (2012) *Guide for REDD-plus negotiators,* Foundation for International Environmental Law and Development, October.

---

11 On Australia's climate change relationships with States in the global south more generally, see Elliott (2011).

Garnaut, R. (2009) 'Climate change and Indonesia: In honour of Panglaykim', *Bulletin of Indonesian Economic Studies* 45: 107.

Government of Indonesia (2011) *Second National Communication to the UNFCCC*, United Nations Framework Convention on Climate Change: http://unfccc.int/ national_reports/ non-annex_i_natcom/items/2979.php.

Government of Indonesia (2013) 'Warsaw's little agreement forms greater deals on global climate change after 2020', press release VI, Indonesian Delegation to COP19/CMP9, 20 December www.indonesiacop19.com/en/press-briefing.

Murdiyarso, D. (2004) 'Implications of the Kyoto Protocol: Indonesia's perspective', *International Review for Environmental Strategies* 5: 145.

Saatchi S. S., N. L. Harris, S. Brown, M. Lefsky, E. T. A. Mitchard, W. Salas et al. (2011) 'Benchmark map of forest carbon stocks in tropical regions across three continents', *Proceedings of the National Academy of Sciences* 101: 9899.

Saul, B., S. Sherwood, J. McAdam, T. Stephens and J. Slezak (2012) *Climate change and Australia: Warming to the global challenge*, Leichhardt: Federation Press.

Saul, B. and T. Stephens (2012) 'Not yet out of the woods: Australia's attempt to regulate illegal timber imports and World Trade Organisation obligations', *Australian International Law Journal* 19: 143.

Takahashi K. and A. Kuriyama (2013) *Market mechanisms country fact sheet: Indonesia*, Institute for Global Environmental Strategies, January: http://enviroscope.iges. or.jp/ modules/envirolib/upload/984/attach/indonesia_final.pdf.

United Nations Framework Convention on Climate Change (UNFCCC) (2012a) *Report of the Conference of the Parties serving as the Meeting of the Parties to the Kyoto Protocol on its seventh session*, Tenth plenary meeting, Durban: 11 December 2011, UN Doc. FCCC/KP/CMP/2011/10/Add.1, published 15 March.

United Nations Framework Convention on Climate Change (UNFCCC) (2012b) *Outcome of the work of the AWG-KP*, Conference of the Parties serving as the meeting of the Parties to the Kyoto Protocol, Eighth session, Doha: 26 November–7 December 2012, UN Doc. FCCC/KP/CMP/2012/L.9, published 8 December 2012.

World Resources Institute (2013) *Climate Analysis Indicators Tool 2.0*, interactive online tool, Washington, DC: World Resources Institute, WRI's Climate Data Explorer: http://cait2.wri.org/wri.

4 The institutional environment for REDD+ in Indonesia

This chapter provides an overview of the institutional and legal framework in Indonesia in which any REDD+ scheme will need to operate. It has two main purposes. The first is to provide background that informs the discussion in later chapters. For this purpose, we introduce Indonesia's decentralisation processes, common jurisdictional disputes and mechanisms for their resolution, the judiciary, traditional communities (*masyarakat hukum adat*, literally 'customary law communities'), and freedom of information. Our second purpose is to discuss how features of the Indonesian legal system are likely to have a bearing on REDD+. In particular, we focus on land tenure, prior informed consent and public participation.

We begin with Indonesia's unsteady transition to democracy from authoritarianism. While the reforms associated with the transition are most welcome from a rule of law perspective, we show that they have made the establishment of REDD+ projects inordinately more difficult than if Indonesia had retained a more centralised political and legal system. This is because key decision-making powers – including about natural resource exploitation and conservation – have been spread among various individuals and institutions, both national and regional. As we shall see in subsequent chapters, difficulties arise because the relative jurisdictions of many of these individuals and institutions are not clearly delineated, including in respect of matters relevant to REDD+, such as the issuance of forestry concessions and licences. Worse, on the whole, the mechanisms in place to help resolve these types of disputes are inadequate and function poorly.

From integralism to fragmentation

Indonesia was an authoritarian State during most of Soeharto's 32 years in power (1966–98) and for the last six years of the rule of Indonesia's first President, Soekarno (1945–65). Components considered essential for democracy (see, for example, Dahl 1971) were absent. National parliamentarians were either selected through rigged elections or directly appointed by the executive. Parliament was, therefore, beholden to the executive and little more than a rubber stamp for government – particularly Presidential – policy

(Schwarz 1994). The government, often with military support and by violent means, strictly confined activities of opposition parties, restricted political freedom (Lubis 1993), controlled the media (Hill 2009) and repressed dissent from civil society (Budiman 1994).

There were also virtually no checks on the exercise of government power. Formally, a Constitution bound the State, but no judicial institution had power to hold the government to account (Lev 1978) and, in any event, most judges lacked independence from government (Butt and Lindsey 2010). Corruption among officials, including the judiciary, was widespread and institutionalised (Pompe 2005). The result was a very strongly centralised and tightly controlled State supported by a legal system that consistently failed citizens but served the government well, providing legal impunity for State actors.

The Soeharto regime often justified the repressive nature of the State by pointing to the importance of political stability (*stabilitas*) to national development (*pembangunan*). These justifications were strengthened by at least two ideological mechanisms. The first was Indonesia's national ideology, *Pancasila* (the 'Five Principles') with which the Indonesian citizenry was indoctrinated, thereby giving it semi-sacred status. One of its five principles is 'national unity'.[1] From the 1980s, this was interpreted in line with the 'integralistic State' concept, under which the State and the people were portrayed as a single entity – an 'organic whole'.[2] Although not explicitly labelled as such, this notion was little more than pure authoritarianism. There was no room – indeed no need – for mechanisms to challenge State action, including through judicial review by an independent judiciary (Lubis 1993). There was also no need to divide power among various government institutions because all government institutions were presumed to have oneness of purpose. The primary function of local governments – provincial, municipal, city and village – was to support and implement national policies and directives.

When Soeharto wrestled power from Soekarno in 1966, Indonesia's economy was in a parlous state. Inflation had reached almost 600 per cent per annum, per capita income was around US\$70 and Indonesia's foreign debt was very high. The Soeharto regime saw increased exploitation of Indonesia's abundant natural resources, including forests, as a means to drive economic

---

1 *Pancasila* (literally 'The Five Principles') embodies a commitment to the following principles:
   1 *ketuhanan yang maha esa* (belief in unitary deity)
   2 *kemanusiaan yang adil dan beradab* (a just and civilised humanity)
   3 *persatuan Indonesia* (unity of Indonesia)
   4 *demokrasi* (democracy)
   5 *keadilan sosial* (social justice).
2 This concept first emerged in 1945 when drafters of Indonesia's first Constitution debated what form the Indonesian State should take after achieving independence. They considered formally adopting the integralistic State and Islamic State models, but chose instead a 'state based on law' (*Rechtsstaat*).

growth, particularly in the timber, pulp and paper industries. Licences over vast swathes of forests were granted, often to Soeharto cronies or conglomerates and without tenders, with little attention to environmental impact and remediation. The regime awarded itself exclusive rights to exploit more than 140 million ha of forests by enacting the 1967 Forestry Law[3] under which concessions were managed almost entirely by the Forestry Ministry (Burgess et al. 2012: 1716). The process became highly centralised, monopolistic and corrupt and the Suharto regime was said to enjoy a significant share of the proceeds (Lindsey 2004).

Since Soeharto's fall, amidst social, political and economic unrest,[4] radical democratic and constitutional reform has taken place, dismantling key components of the 'integralistic State'. Indonesia has been transformed from one of Southeast Asia's most repressive political systems to one of its most free.[5] The concentration of power in the Presidency has, in particular, been dispersed, with much of it allocated to the parliament, which is now democratically elected.[6] Government ministries and other national institutions now enjoy more autonomy than ever before. Provincial and country/city executives and legislatures now have sweeping new lawmaking powers. Judicial institutions are now independent from government. The Supreme Court can and does invalidate administrative decisions and some types of government law. The Constitutional Court has jurisdiction to invalidate statutes enacted by the national parliament that violate the Constitution and regularly exercises that jurisdiction.

This radical devolution of power makes introducing and maintaining any legal infrastructure to support REDD+ far more difficult in Indonesia now than at any time in its history. Presidential support for REDD+ would, during Suharto's reign, have likely ensured its adoption and perhaps even its success. In today's Indonesia, however, the President's voice, although louder than most, is but one in a cacophony of many others – including Indonesia's hundreds of government departments and institutions, whether national, regional, judicial, executive or legislative.

3 Law 5 of 1967 on Forestry.
4 When Soeharto stepped down in May 1998, Indonesia was in the midst of economic and monetary crisis. The economic calamity – a flow-on from the collapse of the Thai baht in July 1997 that had caused many foreign investors to re-evaluate their portfolios in Indonesia – unravelled much of the economic development achieved during Soeharto's time in power. Indonesia lost 13.5 per cent of its GDP in 1997 alone and its currency plummeted from Rp 2000 per US dollar to almost Rp 20,000 by February 1998.
5 See www.freedomhouse.org.
6 Indonesians now vote in more free, fair and highly competitive elections than citizens of 'nearly any other democracy' (Ramage 2007: 136). In 2004 and 2009 more than 150 million citizens voted for two legislatures (their local parliament and the national parliament) and a national-level regional representative body (the *Dewan Perwakilan Daerah*, or DPD). The President and Vice-President are now also directly elected as are the heads of provinces, districts and cities.

Decentralisation and jurisdictional conflicts

In 1999, the year following President Soeharto's resignation, Indonesia embarked on an ambitious programme of decentralistion, or 'regional autonomy' (*otonomi daerah*).[7] Well before Soeharto's departure, the central-ised and authoritarian system he established had become deeply unpopular. Many provinces had long complained that Jakarta's economic, military and bureaucratic controls were excessive. They also protested that the fruits of Indonesia's natural resources, largely located in outer regions, were chan-nelled almost entirely to the centre – particularly to the Soeharto family. Other provinces, such as West Papua and East Timor, claimed on historical grounds that they should never have been brought under the banner of the Indonesian State. Discontent became so strong and sustained that Indonesia faced serious threats of separatism and, ultimately, disintegration. On one view, the only way the central government could retain any power at all was to put autonomy into effect (Ferrazzi 2000).

Under the 1999 decentralisation laws,[8] broad and wide-ranging autonomy (*otonomi yang luas*) was granted to the various levels of regional government – particularly, provinces (*propinsi*), districts (*kabupaten*) and cities (*kota*) – to make their own policies and local laws, usually referred to as *perda* (*peraturan daerah*, or regional regulations), except about matters reserved exclusively for the central government. At time of writing, Indonesia had 34 provinces and almost 500 municipalities and cities.[9]

The 1999 laws were followed by constitutional amendments, passed in 2000. Article 18(1) of the Constitution confirms that Indonesia is to be divided into provinces and that these provinces are to be divided into districts and cities. Each of these three levels of government is to:

- have its own regional government (Article 18(1)) and parliament with elected members (Article 18(3))
- manage and regulate the activities of government, as an expression of autonomy or in assisting the central government (Article 18(2))
- have democratically elected governors (for provinces), regents (for dis-tricts) and mayors (for cities) (Article 18(4)).

Article 18(5) of the Constitution states that regional governments are to have autonomy that is 'as wide as possible' (*otonomi seluas-luasnya*), except over affairs of government reserved by statute for the central government. To fulfil this mandate, Article 18(6) allows the regional government to issue *perda*.

Finally, the two 1999 statutes were replaced by Law 32 of 2004 on Regional Autonomy and Law 33 of 2004 on Fiscal Balance between the Central and

7 This section draws on Butt (2010).
8 Law 22 of 1999 on Regional Autonomy; Law 25 of 1999 on Fiscal Balance between the Central and Regional Governments.
9 See the Home Affairs Ministry website: www.depdagri.go.id.

Regional Governments. At time of writing, these 2004 laws remained in force, although they had been amended.[10]

These local governments have, by exercising these law- and policymaking powers, added great complexity to Indonesia's legal system. They have produced tens of thousands of laws between them and have paid scant regard to national laws that purport to limit their jurisdiction (Ray 2003; Butt and Lindsey 2012). One result has been that two or more levels of government often claim authority over a particular matter and issue inconsistent laws. Disputes between institutions of the same tier of government also regularly arise.

Indonesia's transfer from a polity with strong centralised power to one in which the State's power is markedly fragmented and dispersed has not been accompanied with adequate mechanisms to resolve these overlaps and disputes. The absence of such mechanisms has already frustrated various attempts to regulate aspects of REDD+ and threatened the long-term viability of effective large-scale REDD+ schemes. Many national ministries and local governments are vying for some form of power in relation to forestry and other natural resources in order to obtain revenue, including by issuing concessions. Some of these institutions claim this power on the basis of legitimate legal authority. However, significant problems arise when more than one institution claims the same power under different laws.

The Supreme Court

At the time of Suharto's resignation in 1998, Indonesia's courts were in a dire state. Judges were required to be civil servants and members of the Public Servant Corps (*Korps Pegawai Republik Indonesia*, or KORPRI).[11] This compelled them to support government policy (Asia Watch 1988: 170). In accordance with their civil servant status, judges could be reprimanded under Government Regulation 30 of 1980 on Disciplining Civil Servants. Lev (1978: 55–6) describes some implications of this:

> Indonesian judges, with few exceptions, tend not to have a strong sense of functional independence to begin with. They conceive themselves as *pegawai negeri*, officials, and as such, members of a bureaucratic class to which high status has always attached. One implication of the role of *pegawai negeri* is that it is patrimonially associated with political leadership, to whose will it must always be responsive. It is this as much as anything else that underlies the issue of judicial independence. Whatever the daily effects of the Ministry's responsibility, it is symbolically important

10 Law 32 of 2004 on Regional Autonomy was amended by Law 8 of 2005 (First Amendment) and Law 12 of 2008 (Second Amendment).
11 Presidential Decree 82 of 1971 concerning the Public Servant Corps of the Republic of Indonesia (KORPRI).

as a reminder of the judiciary's conceptually limited autonomy and the direction of its loyalties.

The Ministry of Justice had organisational, administrative and financial control over the lower courts,[12] rendering judges reliant on the government for employment, pay and promotion. Imbued with a sense of loyalty to the government, judges were said to be reluctant to bite the hand that fed them (Lev 1978). It was widely suspected that, in many cases, the government actively used its administrative or organisational leverage to obtain decisions favourable to the government from the judiciary in cases involving the State (see Pompe 2005). The State grossly underfunded many institutions, including the courts – providing them with around only 30 per cent of the amount they required to meet operational expenses (see Pompe 2005). This pushed judges to seek additional payments from litigants, not just to supplement their very low wages, but also to ensure that the courts could continue providing services. Unsurprisingly, the judicial system degenerated into one in which cases were decided in favour of the highest bidder. This had calamitous consequences for judicial competence: for the most part, judges did not need to pay much attention to the law or the evidence presented to them to decide cases.

Since the fall of Soeharto, important structural reforms have been made to Indonesia's judicial system.[13] These reforms seek to provide a legal basis for the transfer of control over the organisational, administrative and financial affairs of the lower courts from the government to the Supreme Court – the so-called 'one roof' (*satu atap*) reforms. The *satu atap* reforms have widely been regarded as successful, at least to the extent that the government is rarely accused of attempting to influence judicial decisions.

The Supreme Court and the courts below it are now unlikely to side predictably or consistently with the government, including in REDD+-related cases, as would have been likely pre-*satu atap*. This is important, because the government, whether national or local, is likely to be a party in many REDD+ cases. Citizens might, for example, challenge, in the administrative courts, decisions to award mining or logging concessions over land already allocated to REDD+ or seek before the Supreme Court the invalidation of a local government spatial plan that is inconsistent with a national plan. They might also seek redress for being refused compensation for foregoing their entitlements to forests under a REDD+ scheme; or pursue a remedy if prior informed consent was not sought and granted or if no public consultation was held prior to land being allocated for REDD+. The State might become

12 Article 11(1) of Law 14 of 1970 on Judicial Power.
13 Law 4 of 2004 on Judicial Power, which replaced a 1970 Law with the same title and subject matter; Law 5 of 2004, which amended Law 14 of 1985 on the Supreme Court; Law 8 of 2004, which significantly amended the General Courts Law (Law 2 of 1986); Law 9 of 2004, which amended Law 5 of 1986 on the Administrative Courts.

involved in other types of judicial proceedings, including if a citizen reports an official for breach of the criminal law, such as conspiracy to engage in commercial logging without permission. Different tiers of government might approach the Supreme Court to resolve jurisdictional conflicts.

However, despite the advances brought by *satu atap*, serious questions remain about the ability of the Supreme Court, and the lower courts it administers, to resolve these and other types of disputes. High levels of judicial corruption and generally low judicial capacity – both problems that emerged during the Soeharto period – remain, even in the Supreme Court itself.

In Chapter 8, we discuss the judicial institutions that are likely to have jurisdiction over REDD+-related cases – the general courts, administrative courts and Supreme Court – and describe some of the obstacles to professional resolution of REDD+ cases. We also discuss whether citizens are likely to be granted standing by these courts to bring cases that might arise from REDD+ projects and the remedies, if any, those courts can offer to those adversely affected by REDD+. Citizens might suffer damage if, for example, land to which they are entitled is allocated for REDD+ without consultation and compensation or if REDD+ is improperly implemented, resulting in loss of anticipated financial or developmental returns. We also consider whether Indonesia's national Ombudsman is likely to be able to assist citizens resolve complaints against the government.

The Constitutional Court

The Constitutional Court, established in 2003, represents one of the most successful post-Suharto era reforms (Butt and Lindsey 2012). It is the first Indonesian court to have powers of judicial or constitutional review – that is, to ensure that the statutes passed by Indonesia's national parliament comply with the Constitution.[14] The Constitutional Court has nine judges who are inaugurated by the President.[15] Three of them are chosen by the Supreme Court, three by the national parliament and three by the President.[16]

By most accounts, the Court has performed its functions with professionalism and integrity unmatched by Indonesia's other judicial institutions, perhaps even in Indonesian legal history (Butt and Lindsey 2010).[17] The Court had, at least under its first and second Chief Justices Jimly Asshiddiqie (2003–2008) and Mahfud (2008–2013), built a deserved reputation for being politically independent, competent, reliable and impartial in its decision making. Its decisions are generally more discursive and reasoned compared with those issued by Indonesia's Supreme and other courts (Butt 2006). Its

---

14 The Court has several additional constitutional functions, outlined in Chapter 9.
15 Article 4(1) of Law 24 of 2003 on the Constitutional Court.
16 Article 24C(3) of the Constitution.
17 With the possible exception of the religious courts, which have a reputation for being corruption free. See Sumner and Lindsey (2010).

decisions have been largely respected by government, even though it lacks the power to enforce them.[18] This is a significant achievement in a political environment in which many politicians are unaccustomed and even openly hostile to subjecting their laws to review. However, in October 2013 the Court's reputation nosedived, after its third Chief Justice, Akil Mochtar, was arrested on corruption charges. In mid 2014, he was sentenced to life imprisonment for receiving bribes in several electoral disputes. Whether the Court can fully regain its reputation for professionalism remains to be seen. Early indications are positive, however. Since the Mochtar scandal there has been no perceptible drop-off in cases being lodged with the Court or widespread reports of noncompliance with its decisions.

The Court has issued decisions invalidating various provisions in statutes regulating natural resources, including forests, land and oil and natural gas. Some of these decisions have significant potential ramifications for the implementation of REDD+ in Indonesia, making the process of allocating land for REDD+ far more complex and difficult. If the national parliament enacts legislation on REDD+ and fails to heed these prior decisions, individuals and non-government organisations that campaign against REDD+ as presently formulated, such as AMAN (the Indigenous Peoples' Alliance of the Archipelago), will surely lodge a constitutional challenge to that legislation. The Constitutional Court is likely to strike down aspects of the legislation, following its previous decisions.

In particular, the cases discussed in Chapter 9 seem to preclude the national government from establishing REDD+ projects without including regional governments in the planning and running of those projects. (These cases appear to hand significant bargaining power to local governments to negotiate better financial returns from REDD+ than they might otherwise have enjoyed.) The decisions also require the government to respect the rights and entitlements of traditional communities over forest areas. Other cases seem to preclude the national government from 'fencing off' forests for an extended period.

Corruption[19]

Indonesia has levels of corruption that are among the highest in the world. Corruption has been prevalent in government throughout most of Indonesia's modern history – including under Soekarno's 'Old Order' (*Orde Lama*) (1947–66) and even during Dutch colonialism. However, corruption increased significantly and became deeply entrenched in government institutions during Soeharto's reign (Robertson-Snape 1999: 592). There are indications, albeit anecdotal, that, since Soeharto's fall, overall corruption has worsened. Regional autonomy is commonly blamed for this because it dispersed power

18 For some exceptions, see Butt and Lindsey (2008).
19 This section draws on Butt (2012).

to local officials, many of whom exploit their office for private gain (Pratikno 2005; Hadiz 2004; Hadiz and Robison 2005), following the example set by Soeharto himself. Corruption is engrained within all tiers of government – national, provincial, city, municipality and village. Many Indonesians institutions, including democratically elected national and regional parliament, and the judiciary, rank among the most corrupt in Asia (Diansyah 2009).

Under Soeharto, corruption did not simply grow out of low government budgets and official salaries and the failure of government to stamp it out. Corruption was part of a 'conscious effort to generate and harvest rents from business (and, to a lesser extent, from individuals) at all levels' (McLeod 2000: 102). In his seminal piece on Soeharto-era corruption, McLeod (2000: 101–2) describes the system that developed as akin to a franchise:

> Just as Soeharto used his position as head of the national government to bestow privileges on selected firms ('cronies', as they have come to be known), so he effectively awarded franchises to other government officials at lower levels to act in similar manner. This included many of his ministers and senior bureaucrats, government administrators at all levels – from provinces down to rural villages – and top executives in the state enterprises and special government bodies ... These franchises were not awarded free of charge, of course: as with orthodox business franchises, there had to be benefits to both franchiser and franchisee.

Widespread corruption in the forestry sector and illegal logging in conservation areas are decades long and notorious problems in Indonesia. Businesses interested in exploration or extraction might bribe officials, whether national or local, to obtain a concession, licence or permit. A bribe might be needed if, for example, the applicant fails to meet one or more of the formal prerequisites to obtain the licence, to prompt the official to consider a legitimate application after some delay or to obtain a letter of recommendation required for a permit (Indrarto et al. 2012: 46). Local and national officials are often accused of granting illegal concessions in return for illicit payments or 'kickbacks' from recipients. They might, for example, award a concession to exploit resources located on or within land designated for another purpose. Officials might also be bribed to turn a blind eye to a breach of licence conditions, such as engaging in activity beyond the territorial scope of the licence or extracting more natural resources than that licence permits. They might also be 'convinced' to change forest zoning designations so that 'legal' concessions to exploit can be awarded over protected forests (Human Rights Watch 2009).

In 2011 Indonesia's leading anti-corruption watchdog, Indonesian Corruption Watch (ICW), released findings indicating that forestry sector corruption in Central and West Kalimantan alone had resulted in State losses over US$1 billion (*Jakarta Post* 2011). ICW also discovered at least seven companies in West Kalimantan had converted forested areas into palm oil plantations and at least 15 companies in Central Kalimantan had illegally managed 211,580 ha of forest between them (*Jakarta Post* 2011). A 2011 Reuter's report claims that Indonesia's Forestry Ministry earns US$15 billion a year in land permit fees and ICW estimates that US$2.3 billion of that comes from illegal logging and kickbacks from improperly issued licences (Fogarty 2011).[20]

It is widely accepted that a significant factor contributing to Indonesia's corruption levels is the so-called 'justice-sector mafia' (*mafia peradilan*) (ICW 2001). This is a reference to police, prosecutors and judges, many of whom are said to extort or take bribes, including when handling corruption cases. For a fee, many of them are said to be willing to use their authority to drop investigations or present sloppy prosecutions. Popular belief has it that most of Indonesia's judges and court officials are willing to accept, or even to extort, bribes from litigants to secure victory in their cases. Former Chief Justices of the Supreme Court have themselves admitted to widespread corruption among the judiciary (Pompe 2005: 414; Inside Indonesia 1997).

Clearly, a 'justice' system that is, on the whole, this corrupt cannot be relied on to professionally handle any type of case, let alone a corruption case. By paying off officials, perpetrators can avoid being effectively investigated, prosecuted or convicted for corruption. According to Assegaf (2002: 130): 'It is obviously impossible to eliminate corruption if the very law enforcement agencies with responsibility for handling corruption cases have become centres of corruption.' Similarly, Goodpaster (2002: 97) observes that:

> Corruption in the legal system eviscerates Indonesia's reform efforts because the system by and large cannot be trusted to render honest decisions, but may be trusted to protect corrupt activities.

In the past decade, however, significant progress appears to have been made in tackling corruption in Indonesia.[21] This, it seems, has been largely achieved by taking from ordinary police, prosecutors and courts the power to handle all corruption cases. Many corruption cases are now investigated and prosecuted by the Anti-Corruption Commission (*Komisi Pemberantasan Korupsi*, or KPK), established in 2003, and all corruption cases are tried in the Anti-corruption Court (*Pengadilan Tindak Pidana Korupsi*, or ACC).

---

20 The Forestry Ministry's management of the so-called reforestation fund (*dana reboisasi*), established during Soeharto's reign to finance the rehabilitation of degraded land and forests, is an infamous example of corruption within the Ministry. Over 20 years almost US$6 billion was paid into this fund by timber concession holders, yet large sums were allegedly used to support politically favoured projects with no connection to deforestation or were directly embezzled by political elites (Barr 2006: 60).

21 Although it seems that, most recently, Indonesia's corruption efforts may have waned. Transparency International's 2012 Corruption Perceptions Index ranks Indonesia at 118 out of 176 countries, giving it a score of 32 out of 100. By contrast, in 2010, Indonesia was ranked 110 out of 176 countries. The lower on the list, the worse the perceived corruption. See www.transparency.org.

The KPK is institutionally independent of government. It can investigate and prosecute most corruption cases itself and can take over corruption investigations and prosecutions from police and prosecutors. It has investigative powers that the police do not. For example, it can wiretap suspects' phones without seeking court approval, freeze bank accounts and issue travel bans. The KPK is also restricted in ways that ordinary police and prosecutors are not. For example, once the KPK names a person as a suspect (*tersangka*) – which is equivalent to 'charging' that suspect – the KPK must proceed to trial, even if subsequent evidence shows that the suspect is innocent. The purpose of this mechanism is to preclude KPK officers being bribed to drop a case before trial (Fenwick 2008).

The ACC – a single court located in Jakarta as a branch of the general court – was established to try cases the KPK prosecutes. Unlike the general courts, where three career judges sit on most panels, the ACC was designed with five-judge panels, comprising two career judges drawn from the general courts and three so-called ad hoc judges. These are legal experts, usually academics, practitioners and retired judges, employed to sit on ACC trials. The rationale for their inclusion was that, because they work outside the existing largely corrupt judiciary, they were less likely to seek or accept bribes to fix the outcomes of cases (Fenwick 2008). Appeals were available to similarly constituted benches in high courts and the Supreme Court.

These innovations were immensely effective, at least if one assesses success by reference to conviction rates. At all levels – at first instance and on appeal – ACC panels had maintained a 100 per cent conviction rate in almost 200 decided cases in cases brought by the KPK. However, in 2009, parliament revised the legislation governing the ACC, requiring that regional ACC courts be established in the capital cities of Indonesia's provinces (see Butt 2012). In December 2010 Indonesia's Supreme Court opened the country's first regional ACCs in Bandung, Surabaya and Semarang to operate alongside the Jakarta ACC. By October 2011 all regional ACC courts had been established. These courts are responsible for adjudicating all corruption and money-laundering cases that occur in the territories of their respective provinces. Previously, a small number of cases involving regional corruption had been tried by the Jakarta anti-corruption court, but most of them were handled by the relevant general court located in the province, city or district.

The establishment of these regional ACCs is of great significance to the handling of corruption in natural resource management in Indonesia, at least outside Jakarta. It is in these courts that the KPK will prosecute forestry-sector corruption that occurs in regional areas. (The KPK has identified many possible indications of corrupt practices in regional forest management, but is yet to prosecute many cases.)

Without an effective system of corruption law enforcement, any REDD+-related funds intended for distribution to beneficiaries might find their way into the pockets of government officials. Whether these regional ACCs will professionally adjudicate cases of alleged corruption in the forestry sector, as an effective REDD+ scheme appears to require, is unclear.

Customary law, participation, consent

One prerequisite to effective REDD+ projects is the identification of any legal, or even cultural, entitlements to land and resources that will, or are likely to, be 'closed off' under the project. This distinction between 'legal' and 'cultural' is important, because many Indonesians, particularly in rural areas, are likely to have no formal legal claim to the land and forests on which they rely for their livelihoods. They are, therefore, unlikely to have land certificates to evidence their ownership or other rights to use the land obtained under the primary State land statute: the 1960 Agrarian Law.

Rather, their entitlement is most probably to be found within their own *adat* or customary law. Most Indonesian rural land and forest is held, and used, under *adat* law. Most of the land likely to be earmarked for REDD+, therefore, will be subject to *adat*, which governs fundamental issues such as who owns and controls the land, what it can be used for and how it can be used.[22]

Under some *adat* systems, land can be held by individuals, but in most it is held communally under the so-called *hak ulayat*: the *adat* community's control of the allocation and use of land, usually decided by village head. *Adat* communities recognise that the effort and capital put into a piece of land by an individual creates something of a personal tie between that person and the land, but the *hak ulayat* is the primary *adat* equivalent of 'ownership' (Fitzpatrick 1997).

For several reasons, *adat* is fundamentally unsuitable for determining entitlements for the purposes of REDD+. First, with some exceptions, there is significant variety between *adat* throughout Indonesia, with *adat* in some places providing for different types of title and land use than in others. In an excellent study on land tenure in Aceh, Dunlop (2009) found very high levels of uncertainty and inconsistency in rules on the use of forest resources, even within short distances. Nevertheless, she did find some consistencies between systems, at least when described with a high level of generality:

- land and forest ownership was largely communal rather than individualised;
- most systems only vaguely distinguished between use and ownership, if at all, with many communities giving permission to individuals or groups to use land for a specified time;
- 'outsiders' were usually prohibited from using community land;
- many systems permitted use of timber harvested from forests for particular purposes, including daily needs, such as cooking, fences, and gardens; and community needs, such as building mosques;

22 For an excellent introduction to *adat* law in Indonesia, see Hooker (1978).

- felling trees to clear land for agriculture and plantations was permitted if necessary for the livelihood of the local community;
- felling of high-value species of wood for sale was permitted, although selling to outsiders was generally prohibited;
- using particular types of felled trees for specified purposes was permitted;
- for environmental reasons particular trees were prohibited from being felled, such if they were located close to water sources (Dunlop 2009: 32–42).

However, despite these apparent consistencies, Dunlop (2009: 10) found:

> [H]igh levels of uncertainty surrounding tenure and a lack of information at the community level about the land over which local communities claim customary rights. Differences of opinions about the application of laws, policies and customary laws also existed both between and within communities, government and academics.

One implication for REDD+ of the general diversity and inconsistency of *adat* is that many different *adat* communities are likely to claim entitlements over any large forest tract identified for REDD+ purposes. Because of the variety of different customary laws adhered to by these communities, all of the various *adat* rules must be identified and respected. In other words, it will be insufficient to identify the *adat* that applies to land use and entitlements in only one area and presume that it applies in other areas of the same forest area earmarked for REDD+. The entitlements under communities' *adat* will, therefore, require separate determination.

A second reason why *adat* is largely unsuitable as a basis for determining entitlements is that ascertaining the precise content of *adat* can be difficult: it is often highly fluid and vague and usually oral. Competing interpretations of customary law entitlements within the same community are commonplace (Pompe 2008).

The final reason we discuss here is the formal position of *adat* within the formal Indonesian legal system. On the one hand, the Indonesian Constitution was amended in 1999–2002 and now provides constitutional protection for *adat* rights. Article 28I(3) provides that cultural identity and traditional rights of *adat* communities are to be respected and protected by the State as human rights. Also, Article 18B(2) states: 'The State recognises and respect customary law communities along with their traditional rights.' A line of recent Constitutional Court decisions appears to give some effect to these rights. Although the precise impact of these decisions is unclear, they are likely to improve protection for *adat* law communities, particularly in forest areas.

Yet despite the formal constitutional lip service paid to *adat* law, its position within the Indonesian legal system has, for the most part, been extremely weak in practice for decades. In particular, State law has usually prevailed over *adat* to the extent of any inconsistency, allowing *adat* to formally and autonomously apply only in the absence of State law. This has, in practice, given *adat* very limited space in which to flourish.

The susceptibility of *adat* to override by laws of the State, particularly in matters of land law, is highlighted in the leading academic works on the Basic Agrarian Law (BAL) of 1960 (Hooker 1978; Fitzpatrick 1997). Even though the BAL declares that Indonesian land law is based on *adat*, the Law itself established a new range of statutory rights that overrode *adat* law and left *adat* with very little remaining autonomous authority. Many of these new land rights were particularly western in nature, and many of them were contrary to principles recognised in many *adat* systems. Most notably, the BAL established as the 'fullest' and 'strongest right' the 'right of ownership' (*hak milik*), which is capable of being registered, transferred and mortgaged. Hooker (1978: 118) argues that this ownership right is similar to western legal notions of land ownership and represents 'a radical departure from traditional [*adat*] views on ownership' represented by *hak ulayat*, described earlier.

However, while mentioned in the BAL, *hak ulayat* land is not protected by it. Indeed, *hak ulayat* land is not even registrable under the BAL. On the contrary, the BAL declares that the exercise of *hak ulayat* must conform with national interests,[23] suggesting that it had obstructed regional development.[24] The New Order government, and the post-Soeharto central and regional governments that followed, have treated *hak ulayat* land as State property, often by reference to the BAL (Fitzpatrick 1997: 186). The result is that *hak ulayat* is extremely vulnerable to being taken over by the State or by those who claim statutory rights over the same land.[25]

In times past, disregard of unregistered community-held claims – notably, *hak ulayat* land – might not have presented insurmountable impediments to REDD+. The State could simply have ignored traditional entitlements to forests and issued concessions and permits as required, quelling discontent with threats of, and perhaps even actual, violence. However, in our view the State is no longer likely able to do this successfully in modern Indonesia. Domestically, Indonesia's Constitutional Court has invalidated legislative provisions that failed to protect traditional forest community rights. It is

---

23 Article 3 of the BAL.
24 See Explanatory Memorandum, part A, section 3, paragraph 2. The Elucidation to Article 5 even states 'it is unjustifiable' for a customary law community to use its *hak ulayat* to block various land concessions if doing so would contravene 'broader interests of the nation and state', such as development.
25 Agrarian Minister/National Land Affairs Head Regulation 5 of 1999 on Guidelines for Resolving Adat Community Hak Ulayat Problems (the 1999 Regulation) leaves it to regional governments to determine and recognise *hak ulayat* in their respective regions by issuing *perda* (Article 5). The 1999 Regulation specifies that *hak ulayat* claims must be supported by a legally defined community that continues to observe *adat* in its daily life and has effective customary law institutions that regulate control and use of *ulayat* land. Even though it provides for mapping of *hak ulayat* (i.e. determining boundaries), it does not provide for registration of *hak ulayat* based on those boundaries.

likely that the Court would do this again if a REDD+ scheme failed to respect those rights. Internationally, resolving these issues of land tenure and entitlement is viewed as a prerequisite to successful REDD+ projects (Wright 2011; Sunderlin et al. 2013). Indeed, land tenure certainty itself is one of the principal REDD+ safeguards. It is also fundamental to the meeting of other safeguards. Without establishing ownership or entitlements, it will be impossible to obtain the prior informed consent of those affected by the project, discussed later; it might also be hard to allocate any financial or other benefits arising from REDD+ to those who are entitled to it; and holding any party to account for failure to meet REDD+ obligations will be more difficult.

Lack of certainty is also likely to lead to increased disputation over entitlements as individuals and communities claim land entitlements to reap any forthcoming benefits of REDD+. It is important to note the dramatic increase in agrarian protest since *reformasi* (Lucas 2003) even before REDD+ was conceived. Many of these disputes have been initiated by communities that felt dispossessed during the Soeharto period, when community land was often acquired with no or low compensation for 'development' projects such as timber and mining concessions. If this is a predictor of future behaviour, then if REDD+ fails to recognise customary land tenure, there is likely to be a very strong backlash, which is likely to threaten or even derail REDD+ projects.

In this context, we emphasise the importance of maintaining the integrity of REDD+. Investors hope to sell carbon credits to individuals and companies wanting to reduce their personal 'net' carbon emissions or 'green their image'. At time of writing, individuals and companies were the only confirmed purchasers for forest carbon credits (Morgan 2011). If traditional land rights and entitlements are ignored as land is allocated for REDD+, and impoverished rural communities face dispossession or exclusion from resources on which they and their ancestors have relied for centuries for their subsistence, then REDD+ projects are likely to suffer reputational damage. This is likely to devalue existing investments and, ultimately, discourage further investments. Concerns about 'cultural genocide' have led the Central Kalimantan chapter of AMAN to call for an 'immediate moratorium on REDD+' in Central Kalimantan (*Herald Sun* 2011). It is also likely that traditional communities will ignore REDD+-based restrictions on land use, leading to significant leakage and, presumably, interruption to or cancellation of any financial benefits REDD+ promises.

Prior informed consent and public participation

Once those with entitlements over land likely to be affected by a proposed REDD+ project are identified, it is critical that their informed consent be obtained prior to that REDD+ project commencing. Obtaining prior informed consent from communities who live in or near a REDD+ site is fundamental – not only for those communities, whose livelihoods might be affected by such projects, but also for the long-term viability of the REDD+ project itself (Sunderlin et al. 2013: 2). For similar reasons, 'public participation' is also recognised as an important safeguard. This is the participation of the community likely to be affected by a proposed REDD+ project, allowing it input into policies and choosing what benefits it would like in return for abstaining from forest exploitation. These might include new jobs for community members (including as forestry wardens), alternative livelihoods (such as production of rattan and sandalwood), gifts or cash (Morgan 2011). If local communities do not participate in decision making about REDD+ and have a say in the nature of their 'compensation', then it is likely that they will continue using forests, possibly thereby releasing carbon stored therein, or simply move to another forest area to resume their activities.

Some public consultations do, in fact, take place (see Indrarto et al. 2012: 80–2). However, a common complaint is that community input is simply ignored, with the government often treating consultations as awareness raising or 'socialisation' (*sosialisasi*) exercises to inform communities about projects and policies that have already been decided. One report described this problem in relation to input provided by the AMAN NGO into a draft Forestry Ministry regulation:

> [E]ven though the requirement for participatory processes in drafting the regulations was technically fulfilled, they did not necessarily affect the outcome, and no clear reasons for the acceptance or rejection of the recommendations have been made available. (Indrarto et al. 2012: 79)

There have been numerous reports of inadequate, or no, engagement with affected local communities about plans to allocate land for REDD+ and of failure to obtain prior informed consent (Morgan 2010).[26] The problem is not the lack of statutory and regulatory rights to be consulted on REDD+. Many relevant national-level laws provide these rights. For example, there is a general right contained in Article 96 of Law 12 of 2011 on Law Making, which gives community members the right to provide written and verbal input in the creation of laws. Article 70 of the 1999 Forestry Law also provides a general guarantee of community participation in forestry matters (Indrarto et al. 2012: 24). It states that:

---

26 Morgan (2011) highlights two REDD projects in which community rights and interests are satisfactorily protected. Projects in the Danau Sentarum area of West Kalimantan and another in Aceh 'use participatory forest management techniques, designing the REDD project on traditional land use patterns within village customary forest areas, and developing access plans with communities based on their traditional lands rights and management practices. Community members are developing access rules and restrictions themselves, and community groups will manage forest blocks. Both of these projects' management plans emphasise community members.'

(1) The community is to participate in development in forestry.
(2) The government must support community participation, through various productive and useful activities.
(3) In the framework of increasing community participation, the central and regional governments can be assisted by a forestry observers' forum.
(4) Further details on Articles 70(1) and 70(2) will be regulated by Government Regulation.

Law 26 of 2007 on Spatial Planning also provides similar rights. Article 65 states that 'Spatial planning is implemented by the government with the involvement of the community.' Article 65(2) specifies that the community's role in spatial planning includes participation in spatial plan formulation, using the space allocated in the plan, and controlling any benefits arising from the use of that space. The Spatial Planning Law even gives those who are disadvantaged by the implementation of the spatial plan a right to claim compensation in the courts (Article 66(1)). Article 66(2) states that the defendant can defend such a claim by proving that no breaches occurred in the implementation of spatial planning. It is arguable that if the community is not consulted about a spatial plan and is ultimately disadvantaged by it, then its members might be able to claim for compensation.[27]

More specifically, community participation is to be sought over proposed changes to forest allocation and functions and the issuance of permits over State forests under Government Regulation 10 of 2010 on Procedures for Changing State Forest Allocation and Functions. Article 30(2) of this Regulation requires that State forest allocation and functions at the provincial level be integrated with spatial plan processes (Article 30(2)), thereby arguably importing the community participation requirements of Article 65 of the Spatial Planning Law. Legally speaking, communities are, therefore, entitled to participate in discussions about whether regional spatial plans should allocate land for REDD+ and, if so, where.[28]

Complaints about lack of public consultation and community participation in key REDD+ decisions and developments are therefore really about the in-practice disregard for these laws by the Indonesian government and perhaps even international donors and investors. This is possible, at least in part, because many laws purporting to require public participation mentioned above do not generally impose sanctions for non-compliance. It also occurs when sanctions themselves are not enforced because local communities do not report breaches to police or other relevant authorities. Of course, for local communities to report breaches, they must be aware of relevant laws

---

27 See, further, Chapter 8.
28 Community participation is also to be sought when forest boundaries are physically demarcated. Article 19(2) of Government Regulation 44 of 2004 provides that stages of determining boundaries include inventorising and resolving rights of third parties along the border and in forest areas and for the drafting of formal recognition by the community near the border on an initial indicative demarcation.

and their rights under them. Unfortunately, many traditional communities, with insufficient access to resources, are not so aware. Even if they do report breaches, they rely on the police or the relevant authorities to investigate those breaches. Widespread corruption in law enforcement agencies makes it likely that those accused of breaches can buy their way out of trouble in any event.

Transparency and accountability

On 30 April 2008 Indonesia's national parliament enacted Indonesia's first 'freedom of information' statute – Law 14 of 2008 on Disclosure of Public Information (the 'FOI Law'). The FOI Law gives rights to Indonesian citizens and legal entities to seek and obtain public information held by public bodies. The Law's starting point is that information held by public bodies is public and must, depending on the nature of the information, be available for disclosure on a procedurally correct request from a citizen or legal entity, proactively disclosed periodically or available at all times. A public body can refuse disclosure only if the requested information falls within one of the categories of excluded information listed in the FOI Law or another law. The FOI Law requires public bodies to nominate or employ information officers to service requests for information. It provides dispute resolution avenues for unsatisfied information seekers – such as mediation, 'non-litigation adjudication' and judicial appeals, including to Indonesia's Supreme Court – as well as criminal penalties for officials who fail to comply with information requests. The Law establishes a central Information Commission and regional information commissions to set public information policies and to help citizens and legal entities obtain information if public bodies refuse to disclose it.

Legally, the FOI Law is a significant advance. Prior to its enactment, most public bodies were not required to disclose information. Disclosure, if it occurred at all, was either voluntary or obtained through media pressure. Members of the public had only a very narrow 'right' under a 2007 Supreme Court Chief Justice Decree to access court-related information, including judicial decisions and case statistics. However, some lawyers complained that they were unable to obtain particular court decisions, despite their requests following procedures set out in the Decree. This reflected the view held by many judges that judicial decisions were confidential and should only be made available to the parties to a dispute.

In Chapter 8, we critically analyse the FOI Law and assess how it has been applied, particularly by the Information Commission – the primary institution established by the Law to resolve disputes between people seeking information and public bodies that are thought to hold that information. We also examine administrative court appeals against Commission decisions. We show that disputes that come before the Information Commission and the Indonesian courts were, at the time of writing, almost always being decided in favour of the person seeking the information. In other words, public bodies are usually compelled to disclose information that they would rather keep within their own ranks. Notably, these public bodies include Indonesia's National Intelligence Agency, the police force, many government departments and even some NGOs. However, we demonstrate that significant legal and institutional impediments remain in the way of an effective freedom of information system. We also discuss two administrative court decisions in which public bodies have appealed Information Commission disclosure orders. These appeals were upheld, with the courts allowing the information to remain secret, but employing highly questionable legal grounds.

Nevertheless, we argue that, on the available evidence, Indonesia's reforms in this area have, on the whole, thus far been largely successful. Of course, this bodes well for those seeking information – including associated contracts, concessions and licences and even government policies and plans – about REDD+ projects that have adversely affected them. This information might assist them to bring, and ultimately prove, a claim for breach of their legal entitlements before one of Indonesia's courts, discussed in Chapters 8 and 9.

Conclusion

This chapter has examined the institutional framework in which REDD+ needs to operate in the Indonesian context. The Indonesian legal system has undergone profound changes in recent decades, with the three arms of government now operating within a defined constitutional and legal framework that bears many of the hallmarks of the rule of law. Parliament is a forum for vigorous democratic debate and is active in passing legislation on a range of topics, including forestry. The courts have become, with some exceptions, independent and impartial arbiters of disputes and shown a willingness to subject the parliament and the executive to judicial review and to uphold the rights of Indonesian citizens. There have also been important steps in making the Indonesian government more transparent and accountable, as seen in the adoption of the FOI Law.

Despite these important developments key challenges remain for successfully implementing REDD+, and indeed any other major legislative reform in Indonesia. First, the policy of decentralisation has meant that legal and political power has become highly devolved and fragmented. This presents great challenges for any successful implementation of a national REDD+ system as the complexity of legal and political frameworks at provincial and country/city levels makes it difficult to design and give effect to a one-size-fits-all REDD+ mechanism. Yet for any market in REDD+ credits to be effective and credible it will need to operate in as clear and straightforward a way as possible. The second major challenge facing REDD+ in Indonesia, which relates to the fragmentation of political and legal authority, is endemic corruption at all levels of Indonesian government. However, we note that there have been major improvements in curbing corruption in recent years, providing good reasons to believe that REDD+ can be made to function effectively if there is appropriate oversight of all government decision makers involved in its administration.

References

Asia Watch (1988) *Human rights concerns in Indonesia and East Timor*, New York: Asia Watch Committee.

Assegaf, I. (2002) 'Legends of the fall: An institutional analysis of Indonesian law enforcement agencies combating corruption' in T. Lindsey and H. W. Dick (eds) *Corruption in Asia: Rethinking the governance paradigm*, Annandale: Federation Press.

Barr, C., I.A.P. Resosudarmo, A. Dermawan, J.F. McCarthy and M. Moeliono (2006) *Decentralization of forest administration in Indonesia: Implications for forest sustainability, economic development and community livelihoods*, Bogor, Indonesia: Center for International Forestry Research (CIFOR).

Budiman, A. (1994) *State and civil society in Indonesia*, Clayton: Centre of Southeast Asian Studies.

Burgess, R., M. Hansen, B. A. Olken, P. Potapov and S. Sieber (2012) 'Political economy of deforestation in the tropics' *Quarterly Journal of Economics* 127(4): 1707.

Butt, S. (2006) 'Judicial review in Indonesia: Between civil law and accountability? A study of Constitutional Court Decisions 2003–2005', PhD dissertation, Law Faculty, Melbourne University.

Butt, S. (2010) 'Regional autonomy and the proliferation of Perda in Indonesia: An assessment of bureaucratic and judicial review mechanisms' *Sydney Law Review* 32(2): 177.

Butt, S. (2012) *Corruption and law in Indonesia*, London: Routledge.

Butt, S. and T. Lindsey (2008) 'Economic reform when the Constitution matters: Indonesia's Constitutional Court and Article 33' *Bulletin of Indonesian Economic Studies* 44(2): 239.

Butt, S. and T. Lindsey (2010) 'Judicial Mafia: The courts and state illegality in Indonesia', in G. Van Klinken and E. Aspinall (eds) *The State and illegality in Indonesia*, Leiden: KITLV Press.

Butt, S. and T. Lindsey (2012) *The Indonesian Constitution: A contextual analysis*, Oxford: Hart.

Dahl, R. A. (1971) *Polyarchy: Participation and opposition*, New Haven, NJ: Yale University Press.

Diansyah, F. (2009) *Weakening of Corruption Eradication Commission (KPK) in Indonesia: Independent Report*, Independent Report, Doha, Qatar: Indonesia Corruption Watch/ National Coalition of Indonesia for Anticorruption.

Dunlop, J. (2009) *REDD, tenure and local communities: A study from Aceh, Indonesia*, Rome: International Development Law Organisation.

Fenwick, S. (2008) 'Measuring up? Indonesia's Anti-Corruption Commission and the new corruption agenda' in T. Lindsey (ed.) *Indonesia: Law and society*, 2nd edn, Annandale: Federation Press.

Ferrazzi, G. (2000) 'Using the "F" word: Federalism in Indonesia's decentralisation discourse', *Publius: The Journal of Federalism* 30(2): 63.

Fitzpatrick, D. (1997) 'Disputes and pluralism in modern Indonesian land law', *Yale Journal of International Law* 22(1): 171.

Fogarty, D. (2011) 'Special report: How Indonesia hurt its climate change project', *Reuters*, 16 August.

Goodpaster, G. (2002) 'Reflections on corruption in Indonesia', in T. Lindsey and H. W. Dick (eds) *Corruption in Asia: Rethinking the governance paradigm*, AnnandaleFederation Press.

Hadiz, V. (2004) 'The rise of neo-Third Worldism? The Indonesian trajectory and the consolidation of illiberal democracy', *Third World Quarterly* 25(1): 55.

Hadiz, V. and R. Robison (2005) 'Neo-liberal reforms and illiberal consolidations: The Indonesian paradox', *Journal of Development Studies* 41(2): 220.

Herald Sun (2011) 'Indigenous Indonesians slam forest scheme', 22 June.

Hill, D. (2009) *Politics and the media in 21st century Indonesia: Decade of democracy*, London: Routledge.

Hooker, M. B. (1978) *Adat law in modern Indonesia*, New York: Oxford University Press.

Human Rights Watch (2009) *'Wild money': The human rights consequences of illegal logging and corruption in Indonesia's forestry sector*, New York: Human Rights Watch.

Indonesian Corruption Watch (ICW) (2001) *Menyingkap tabir mafia peradilan (Hasil Monitoring Peradilan ICW)*, Jakarta: Indonesian Corruption Watch.

Indrarto, G. B., P. Murharjanti, J. Khatarina, I. Pulungan, F. Ivalerina, P. Justitia et al. (2012) *The context of REDD+ in Indonesia: Drivers, agents and institutions*, Bogor, Indonesia: CIFOR and ICEL.

Inside Indonesia (1997) 'Courting corruption', vol. 49 (January–March).

Jakarta Post (2011) 'Illegal forestry in Kalimantan can cost the state 1bn', 27 September.

Lev, D. S. (1978) 'Judicial authority and the struggle for an Indonesian Rechsstaat', *Law and Society Review* 13: 37.

Lindsey, T. (2004) 'Indonesia: Devaluing Asian values, rewriting rule of law' in R. P. Peerenboom (ed.) *Asian discourses of rule of law*, London: RoutledgeCurzon.

Lubis, T. (1993) *In search of human rights: Legal-political dilemmas of Indonesia's New Order, 1966–1990*, Jakarta: PT Gramedia Pustaka Utama.

Lucas, A. (2003) 'The State, the people, and their mediators: The struggle over agrarian law reform in post-New Order Indonesia', *Indonesia* 76: 87.

McLeod, R. (2000) 'Soeharto's Indonesia: A better class of corruption', *Agenda* 7(2).

Morgan, B. (2010) 'REDD at the community level: Community engagement and carbon conservation in Indonesia's forests', unpublished Master's thesis.

Morgan, B. (2011) 'Community engagement: Don't ignore REDD's impacts on communities!', *Inside Indonesia* 105: July–September.

Pompe, S. (2005) *The Indonesian Supreme Court: A study of institutional collapse*, Ithaca, NY: Southeast Asia Program, Cornell University (Studies on Southeast Asia).

Pompe, S. (2008) 'Between crime and custom: Extra-marital sex in modern Indonesian law' in T. Lindsey (ed.) *Indonesia: Law and society*, 2nd edn, Annandale: Federation Press.

Pratikno (2005) 'Exercising freedom: Local autonomy and democracy in Indonesia (1999–2001)', in P. Sulistiyanto, M. Erb and C. Faucher (eds) *Regionalism in post-Suharto Indonesia*, New York: RoutledgeCurzon.

Ramage, D. (2007) 'Indonesia: Democracy first, good governance later', *Southeast Asian Affairs 2007*: 135.

Ray, D. (2003) 'Decentralization, regulatory reform, and the business climate', in *Decentralization, regulatory reform, and the business climate*, Jakarta Indonesia: Partnership for Economic Growth.

Robertson-Snape, F. (1999) 'Corruption, collusion and nepotism in Indonesia', *Third World Quarterly* 20: 589.

Schwarz, A. (1994) *A nation in waiting: Indonesia in the 1990s*, Boulder, CO: Westview Press.

Sumner, C. and Lindsey, T. (2010) *Courting reform: Indonesia's Islamic courts and justice for the poor*, Sydney: Lowy Institute.

Sunderlin, W. D., A. M. Larson, A.E. Duchelle, I. A. P. Resosudarmo, T. B. Huynh, A. Awono et al. (2013) 'How are REDD+ proponents addressing tenure problems? Evidence from Brazil, Cameroon, Tanzania, Indonesia, and Vietnam', *World Development* (2013) (in press, copy of file with author).

Wright, G. (2011) 'Indigenous people and customary land ownership under domestic REDD+ frameworks: A case study of Indonesia', *Law, Environment and Development Journal* 7(2); http://www.lead-journal.org/content/11117.pdf.

5 Implementing international law in Indonesian law

The development of REDD+ has taken place primarily on the plane of international law and policy, under the umbrella of the United Nations Framework Convention on Climate Change (UNFCCC). This has resulted in the emergence of international rules and standards for REDD+ that require implementation within domestic legal systems of the parties to the UNFCCC and the Kyoto Protocol. Without such implementation through binding measures that accord with international standards, there is little chance that any national-scale programme for REDD+ in Indonesia will be successful. Before turning in subsequent chapters to examine the national framework for REDD+ in Indonesia in detail, this chapter provides an overview and analysis of the ambivalent position of international law in the Indonesian legal system and suggests ways in which the situation may be clarified so that Indonesia's international legal commitments on REDD+ (and in other fields) may be given a firm legal basis.[1]

Indonesia and international law

Indonesia, the world's fourth most populous country, is an important participant in international and regional affairs. It has joined many treaty regimes, including the UNFCCC, Kyoto Protocols and instruments in many other fields and is expected soon to ratify more, including the Rome Statute for the International Criminal Court (Saragih and Aritong 2013; UN Human Rights Council 2012). By 2013 Indonesia had served three terms on the United Nations (UN) Human Rights Council and in 2011 chaired the Association of Southeast Asian Nations (ASEAN).

---

1 This chapter draws on Butt (2014). These include the key human rights treaties, such as the International Covenant on Civil and Political Rights (adopted 16 December 1966, entered into force 23 March 1976) 999 UNTS 171 (ICCPR) various International Labour Organisation (ILO) sponsored conventions, such as the Convention concerning Minimum Age for Admission to Employment (adopted 26 June 1973, entered into force 19 June 1976) ILO C138, the Agreement establishing the World Trade Organization (1 January 1995): http://www.wto.org/english/res_e/booksp_e/agrmntseries1_wto_e.pdf and numerous bilateral free trade agreements and extradition treaties.

Many commentators have criticised Indonesia for failing to comply with some of the international agreements it has ratified (Jetschke 2011; Colbran 2010; Arnold 2009; Eldridge 2002; Juwana 2005; Butt 2002; Butt and Lindsey 2005, 2006). In this chapter, we focus on one factor that explains this mixed record of compliance: the absence of rules specifying how international law enters into force in the Indonesian legal system. This has obvious relevance to the implementation of REDD+ in Indonesia, given that any internationally linked scheme will require the adoption of global rules and standards into Indonesian law.

As long as they comply with their international obligations, it is a matter for individual States to decide how international legal obligations and rights are received into their own domestic legal systems (Cassese 1986). Some States take a 'monist' approach to treaties and/or customary international law (by which international law is automatically part of domestic law) and others a 'dualist' approach (by which international law must be subject to a process of transformation into domestic law). For example, some States provide that, once ratified, international agreements override domestic law to the extent of any inconsistency. Others deny legal effect to international agreements until they are 'transformed' into domestic law. Yet others allow so-called 'self-executing' treaties to be automatically applied (Triggs 2010). Many countries, but not all, allow customary international law to operate automatically in their domestic legal systems.

Indonesia has made no explicit choice about how international law enters domestic law. It is at least arguable that at the time of Indonesia's independence in 1945, it adopted a partly monist approach. The Constitution of 1945 declared that all existing laws and institutions remained in force, until they are replaced by laws and institutions made by the independent State.[2] Indonesia thereby inherited Dutch laws, institutions and the civil law tradition. The Netherlands is well known for being at least 'moderately monistic' (Alkema 2010) since the beginning of the 20th century (Hoof 1996: 195–6; Kartasapoetra and Kartaputra 1984: 13; Schermers 1987).

However, over the intervening decades the situation has been significantly obscured. The Constitution now in force, and Indonesian legislation concerning international agreements, focus on the processes of entering into and negotiating treaties and say little, if anything, about the status of treaties in Indonesian law. Indonesia's Constitution, for example, states in Article 11 that:

(1) The President, with the approval of the National Parliament, declares war and peace and creates agreements with other nations.

(2) When creating international agreements that give rise to consequences that are broad and fundamental to the life of the people,

---

2 Constitution of 1945, Transitional Provisions, Part II; reaffirmed in Government Regulation 2 of 1945.

create financial burdens for the State and/or require amendments to legislation or the enactment of new legislation, the President must obtain the agreement of the National Parliament.

(3) Further provisions on international agreements are to be regulated by statute.

Indonesia's Law on International Agreements, enacted in 2000,[3] determines who can negotiate and sign treaties on Indonesia's behalf and how treaties are ratified under Indonesian law.[4] Prior to this statute, the only additional guidance on the ratification of treaties was provided, rather unusually, in a letter by Indonesia's first President, Soekarno, addressed to the Parliamentary Chairperson (Manan 1985: 104–5). Titled 'Creating Agreements with Other Countries', it said that, in the opinion of the government, Article 11 of the Constitution did not apply to all types of agreement with foreign States. Rather, the letter sought to confine the meaning of 'agreements' to only 'the most important (*terpenting*) agreements' concerning political issues that could affect alliances and State territory; important economic issues, technical assistance or finance; or matters that must be regulated by statute under Indonesian law.

The significant uncertainty and confusion about whether the rules contained in international treaties ratified by Indonesia automatically form part of Indonesian law has serious ramifications for Indonesia's effective compliance with any obligations it assumes under a scheme for REDD+ that is grounded in a binding international legal instrument.

Indonesian legal scholarship

In the absence of a clear legal framework for determining how international law, including international climate change law, can form part of Indonesian law, we may turn to commentary by legal scholars within Indonesia and to the jurisprudence of Indonesia's superior courts.

That the work of highly respected legal scholars and commentators can shape the development of the law is an accepted and well-recognised practice in Indonesia and in many other civil law countries. Indonesian scholarly discourse about the position of international law within the Indonesian legal system is underdeveloped (Sidik 2004: 229). Many Indonesian international

3 Law 24 of 2000.
4 The Law requires that treaties be ratified either by statute or Presidential Decree, depending on the subject matter of the treaty (Article 9(1)). National parliamentary approval is required if the treaty relates to politics, peace, defence and State security; alterations to or delimitation of the territory of Indonesia; sovereignty or sovereign rights; human rights and the environment; the creation of new legal norms; and foreign loans or aid (Article 10). A Presidential Decree can ratify umbrella agreements concerning cooperation in science, technology, economics, trade, culture, commercial maritime affairs, avoiding double taxation and protection of investment (Article 11).

law texts address the issue, but most of them, like the Constitution and Law on International Agreements, focus on the technical aspects of treaty making, such as negotiation, signature and ratification (Sunaryati 2009: 9–16; Wayan 1987: 128). Many also outline monist and dualist theories and the differences between them, but only some then consider which of these prevails in Indonesia (Boer 2000: 50; Sefriani 2011: 86; Atmasasmita 2000: 488).

Most authors identify two reasons why Indonesia is fully or partially monist. The first is the work of Mochtar Kusumaatmadja. Formerly Dean of the Faculty of Law, University of Padjadjaran (Bandung), Indonesian Justice Minister (1973–78) and Indonesian Foreign Minister (1978–83), Kusumaatmadja is a prominent and widely respected proponent of monism and the lead author of Indonesia's foremost international law text, *Pengantar Hukum Internasional*. Kusumaatmadja observes that, while most countries specify the position of international law in their domestic laws or even their constitutions, Indonesia does not. However, Kusumaatmadja argues that this does not necessarily mean that international law is inferior to domestic law in Indonesia (Kusumaatmadja 2003: 88–9). Pointing to Indonesia's European legal heritage, he concludes that Indonesia is monist (Kusumaatmadja 2003: 92). Even in the absence of formal ratification or implementing regulations, he states, 'we should consider ourselves bound by treaties and conventions approved (*disahkan*) by Indonesia' (Kusumaatmadja 2003: 92).

Kusumaatmadja's influence in legal circles should not be underestimated. According to Agusman, who served as Director of Economic and Social-Cultural Treaties at the Foreign Affairs Ministry (2006–10), Law 24 did not seek to clarify the status of international treaties ratified by Indonesia within domestic law because drafters and the Indonesian Foreign Ministry assumed that Indonesia was monist, following Kusumaatmadja's views (Agusman 2010: 104; Agusman 2007: 490). Agusman, whose work provides perhaps the most detailed examination of whether Indonesia follows a monist or dualist approach, lists additional indications of monism (Agusman 2010: 39; Agusman 2007: 490), but not all of these are convincing as they refer to legislation such as Law 39 of 1999 on Human Rights that actually provides for legislative implementation of international law, rather than recognising its automatic application.

Given the paucity of evidence for monism provided in the literature, it is unsurprising that most other Indonesian scholars conclude that Indonesia is dualist. Many of these refer to the example of the 1982 UN Convention on the Law of the Sea (UNCLOS). Indonesia's parliament formally ratified the Convention by Law 17 of 1985. However, this ratification did not disturb Law 4 of 1960 on Indonesian Waters, which remained in force for more than 10 years after the ratification (Aziz and Yusran 2011).[5]

---

5 According to Aziz and Yusran (2011: 10): 'It was law No. 6/1996 (the implementing legislation of UNCLOS), and not Law 17 of 1985 (instrument of ratification of UNCLOS) which replaced Law 4 of 1960 on Indonesian Waters.'

From this brief review, we can extrapolate principles for how Indonesia addresses international legal obligations, including rules relating to REDD+ adopted under the UNFCCC. It would appear that in practice, ratification of a treaty will, in itself, be insufficient to render an international agreement enforceable in Indonesia. Hence, neither the UNFCCC nor the Kyoto Protocol, nor any supplementary agreement or understanding on REDD+ or other topics, will be legally enforceable in Indonesia without steps taken by domestic legal institutions. The treaty's principles, rights and obligations, and perhaps even a translation of the treaty provisions themselves, need to be included in an Indonesian domestic law (Juwana 2010: 97–8).

However, much appears to depend on the precise nature of the international agreement. On one hand, if an international agreement introduces new legal concepts or contradicts pre-existing Indonesian law, then most police, prosecutors, public servants and even judges will not apply the international text directly without 'transformation' into Indonesian law. Many of the treaties falling into this category are multilateral and well known, at least among Indonesia's legal community, and failure to implement them often leads to calls for the government to issue domestic implementing laws. On the other hand, treaties with more specific and detailed subject matter are, in fact, likely to be routinely followed by those to whom they pertain as if they were binding, particularly if there is no relevant contradictory Indonesian law. Many of these are bilateral and largely uncontroversial and they garner relatively little public attention.

It is quite possible, then, that treaties ratified by Indonesia might become formally binding under Indonesian law but will, as a practical matter, lie dormant until their principles are 'picked up' in a domestic law. This approach appears to be consistent with the view of Kusumaatmadja, discussed earlier. He admits that even though the international agreements that Indonesia has joined are formally binding, if those agreements require changes to domestic law, they will often go unheeded by officials until those changes are made. Kusumaatmadja accepts that 'enactment is absolutely necessary if, for example, it requires changes to national statutes that directly touch on the rights of citizens as individuals' (2003: 94). He reasons that new laws or amendments will be required if, for example, international law creates offences not previously known in Indonesian criminal law (2003: 94).

Indonesian judicial treatment of international law

Supporting the argument that Indonesia follows a monist approach, or a variation of it, is the scope that some Indonesian courts have given to international law in their decision making. The Indonesian Supreme Court – the final court of appeal (or *cassation*) for most types of dispute in Indonesia – has directly enforced international law in at least two cases. (There may well be additional examples, but the Supreme Court has published only a small portion of its decisions.) The Constitutional Court, which has powers of constitutional review (see Butt and Lindsey 2012), makes regular reference to international law in its decisions. Although, to our knowledge, the Constitutional Court has not used international law as a source of law independent of domestic sources, it has in some cases treated international law as a highly persuasive guide when interpreting provisions of Indonesia's Constitution. In this way, the Court has allowed international law to permeate Indonesian law.

The Supreme Court

The Indonesian Supreme Court appears to have applied international law in at least two instances. The first was a Supreme Court Directive issued in 2006, in which the Court applied the 1961 Vienna Convention on Diplomatic Relations. The Convention had been formally ratified by Indonesia's national parliament through Law 1 of 1982, but the Convention's provisions had not yet been transformed into national law. Yet in its Directive, the Indonesian Supreme Court applied the diplomatic community principle in Article 31 of the Convention to a domestic land dispute involving the Saudi Arabian embassy in Indonesia (Agusman 2007: 492).

The second instance is the *Landslide* case.[6] This case began in the District Court of Bandung in 2003. The applicants were victims of a landslide in West Java. They filed a class action against Perhutani (a state-owned forestry company) arguing that the forest area had been mismanaged, causing the landslide; and against the government, arguing that the government failed to monitor Perhutani's activities. The defendants' response was that a natural disaster had caused the landslide. Presented with conflicting testimony, the Court found that there was scientific uncertainty about the landslide's exact cause. To resolve the case, the Court resorted to the precautionary principle adopted in Principle 15 of the Rio Declaration, acknowledging that the principle had not yet been adopted in Indonesian environmental law.[7] The Court found the defendants strictly liable and ordered them to pay compensation. An appeal by the government to the provincial High Court was rejected.

The government appealed to the Supreme Court in Jakarta. One ground for appeal was that the lower courts had been wrong to apply the precautionary principle because the Indonesian government had not ratified the Rio Declaration and the precautionary principle had not been adopted in Indonesian law.[8]

The Supreme Court rejected this argument, holding that:

> [T]he [lower court] judges did not erroneously apply the law by adopting rules of international law. The application of the precautionary principle

6 Our description of this case draws from Wibisana (2011).
7 In particular, Law 23 of 1997 on Environmental Management.
8 Supreme Court Decision 1794-K-PDT-2004, 58.

in environmental law was to fill a legal vacuum … the view of the cassation applicant that Article 1365 of the Civil Code could be applied in this case cannot be justified because the enforcement of environmental law is to be performed by the standards of international law. National judges can use rules of international law if they view it as jus cogens.[9]

The Supreme Court's reference to 'a legal vacuum' in this passage is likely a reference to a provision contained in the numerous judiciary laws (*undang-undang pokok kekuasaan kehakiman*) since at least 1970.[10] Article 10 of Indonesia's Judiciary Law (Law 48 of 2009) is the current iteration. It prohibits courts from 'refusing to examine, adjudicate and decide cases brought before them on the basis that the law does not exist or is not clear. Rather, [they are] required to examine and adjudicate them'. If no Indonesian law applies to a dispute before a judge, this provision appears to provide an avenue through which he or she could apply international law directly to the case.

The Constitutional Court

The Constitutional Court has referred to international law in many of its judicial review decisions. However, in its decisions the Constitutional Court has explained neither the circumstances in which it will use international law nor how it will use international law principles. The Constitutional Court has used international law primarily to help it interpret the Indonesian Constitution and Indonesian laws. However, the weight the Court has given to international law appears to be inconsistent from case to case and the Court has not yet explained these inconsistencies.

There are three approaches to international law in the Court's decisions. The first – the weak-use approach – sees the Court refusing to use international law as a reference point, and ignoring or dismissing out-of-hand arguments based on international law from the parties.[11] This approach is encapsulated in the statements of former Justice Roestandi, who warned against over-reliance on international law in several dissents. Although political and international law developments might be relevant in some cases, he emphasised that the Constitution was the highest source of validity for statutes and

---

9 Supreme Court Decision 1794-K-PDT-2004, 85. Indonesia's lower courts have also sought to apply international law, but in limited circumstances. Wibisana (2011) refers to *Wiwiek Awiati v. Minister of Agriculture* (District Administrative Court of Jakarta Decision 71/G.TUN/2001/PTUN-JKT, 2001) and *Wiwiek Awiati v. Minister of Agriculture* (Administrative Court of Appeals of Jakarta Decision 120/2001/Bd.071/G.TUN/2001/PTTUN-JKT, 2002), in which lower courts applied the precautionary principle.

10 See, e.g., Law 14 of 1970 and Law 4 of 2004.

11 See, e.g., *KPK Law* case (Constitutional Court Decision 37-39/PUU/VIII/2010 reviewing Law 30 of 2002 on the Corruption Eradication Commission) 25-6; *Pornography Law* case (Constitutional Court Decision 10-17-23/PUU-VI/2009 reviewing Law 44 of 2008 on Pornography) 136–7. See Zhang (2010).

trumps international law.[12] As Roestandi (2004: 51) wrote extramurally, '[m]y task as a constitutional court judge is to review the constitutionality of a statute as against the Constitution, not to review the Constitution against international law.'

In some of these cases, the Court has explicitly rejected international norms, even though the Court may have, in fact, been influenced by them. In the *Children's Court Law* case,[13] for example, the applicants raised several international law arguments to challenge the validity of statutory provisions that made the age of criminal culpability eight years old for some offences. In its decision, the Court lifted that age to 12, which was in line with the UN Committee on the Rights of the Child. However, the Court emphasised that in adopting this as the age of criminal responsibility, it was not using 'these instruments and recommendations ... [as] a gauge to assess the constitutionality of the age of responsibility for children'.[14] Similar sentiments were expressed by Arief Hidayat, who was appointed to the Court in 2013. During his 'fit and proper' test before the national parliament, he said that 'Indonesia should implement human rights appropriate to the "local context" instead of unconditionally adopting the standards of the United Nations' (Parlina 2013).

The Court's second approach gives more credence to international law, but attributes no real influence to it. Under this approach, the Court refers to international law, but only to support a decision the Court seems to have already arrived at, based on an interpretation of the Constitution and Indonesian law that the Court claims as its own.

The third approach sees the Court relying quite heavily on international legal principles and interpretations from international bodies to help it construe the Indonesian Constitution, the statute being reviewed in the case, or both.[15] This approach is well expressed by one judge in a case involving employment rights: '[I]n order to understand the right to work' in the Constitution, 'it is best to carefully study' various rights in international labour conventions.[16] What appears to distinguish this category of use from the second approach is that the Court might not have arrived at its decision without using international law. Worryingly, however, in some important

---

12 *Soares* case (Constitutional Court Decision 065/2004 reviewing Law 26 of 2000 on the Human Rights Court) 63–4; *PKI* case (Constitutional Court Decision 011-017/2003 reviewing Law 12 of 2003 on General Elections for Members of the DPR, DPD and DPRD) 40–1.

13 Constitutional Court Decision 1/PUU-VIII/2010 reviewing Law 3 of 1997 on Children's Courts.

14 *Ibid.*, 151.

15 For examples of cases in which the Court extensively refers to international treaties, see the *Water Resources Law* case (Constitutional Court Decision 058-059-060-063/PUU-II/2004 and 008/PUU-III/2005 reviewing Law 7 of 2004 on Water Resources) 486; *Abdurrahman Wahid* case (Constitutional Court Decision 008/PUU-II/2004 reviewing Law 23 of 2003 on General Election of President and Vice President).

16 Namely Article 2 of the ICCPR and Article 23 of the Universal Declaration of Human Rights (adopted 10 December 1948) UNGA Res 217 A(III): *Overseas Workers* case No. 1 (Constitutional Court Decision 019-020/PUU-III-2005 reviewing Law 39 of 2004 on Placement and Protection of Indonesian Overseas Workers), per Natabaya J.

cases the Court appears to have misunderstood the international law it used to interpret the Constitution, leading to outcomes that might perplex some international lawyers (Butt, 2014).

Conclusion

While there remains much uncertainty, reasons exist to suggest that Indonesia is a monist jurisdiction in a formal legal sense, at least in part. Some treaties appear to be automatically applied, particularly bilateral treaties that concern specific matters on which Indonesia pledges to cooperate. Also, some Indonesian courts appear amenable to applying international law in their decisions, although reference to international law has been primarily used as an aid to interpretation in constitutional matters. In the available cases, the Supreme Court has used international law only as an aid to fill the gaps in Indonesian law, not to override it.

However, in practice, the system appears to be primarily dualist: many government officials will not act on international norms until they are transformed into Indonesian law. The result is that Indonesia does not, in fact, comply with many treaties until domestic laws are issued to bring those treaties into effect. The 'lag' between ratification and transformation can be many months or even years, given Indonesia's notoriously slow lawmaking processes. Some treaty obligations have simply not been complied with at all (Juwana 2010). Even though in some ways Indonesia's system might appear to follow the 'self-executing treaty' distinction commonly associated with the United States (Triggs 2010), the government and the courts have provided no guidance on how the system should or does work. The result of all this appears to be significant uncertainty about the effect of ratification of any given treaty.

This uncertainty can be and is used by the Indonesian government to its advantage. To the international community, it can claim that ratification will automatically bring treaties into domestic Indonesian law. If challenged, Indonesian officials – some no doubt with a genuinely held belief – can point to Kusumaatmadja and Indonesia's Dutch legal heritage. This makes deflecting international criticism for non-compliance much easier than if Indonesia were openly dualist. However, the government can evade the demands of a domestic audience to apply or enforce a treaty by pointing to the absence of 'transformation'.

Applying the analysis in this chapter to the context of REDD+, it would appear that a secure legislative basis is both desirable and necessary. First, the global 'rules' regarding REDD+ are, at this stage, at a technical, standards-based level rather than being formally binding under the UNFCCC. They would therefore not be considered treaty-based obligations that could be legally binding without implementation. Second, as the implementation of REDD+ requires a host of changes to property rights and the regulation of land use, with direct impacts on the rights of Indonesian citizens, even if REDD+ emerges in a developed form that is comparable with the Clean

Development Mechanism or emissions trading under the Kyoto Protocol, domestic legislative transformation will still be needed.

References

Agusman, D. D. (2007) 'Status hukum perjanjian internasional dalam hukum nasional RI: Tinjauan dari perspektif praktik Indonesia', *Indonesian Journal of International Law* 5(3): 488.

Agusman, D. D. (2010) *Hukum perjanjian internasional: kajian teori dan praktik Indonesia*, Bandung: Refika Aditama.

Alkema, E. A. (2010) 'International law in domestic systems', *Electronic Journal of Comparative Law* 14(3): 1.

Arnold, L. L. (2009) 'Acting locally, thinking globally? The relationship between decentralisation in Indonesia and international human rights', *Journal of East Asia & International Law* 2: 177.

Atmasasmita, R. (2000) *Pengantar hukum pidana internasional*, Bandung: Refika Aditama.

Aziz, S. and R. Yusran (2011) 'CIL research project on international maritime crimes: Indonesia's country report', Centre for International Law, National University of Singapore.

Boer, M. (2000) *Hukum internasional: Pengertian, peranan dan fungsi dalam era dinamika global*, 1st edn, Alumni.

Butt, S. (2002) 'Intellectual property law in Indonesia: A problematic legal transplant', *European Intellectual Property Review* 24(9): 429.

Butt, S. (2014) 'The position of international law within the Indonesian legal system', *Emory International Law Review* 28(1): 1.

Butt, S. and T. Lindsey (2005) 'TRIPs and intellectual property law reform in Indonesia: Why injunctions aren't stopping piracy', *Harvard Asia Pacific Review* 8(2): 14.

Butt, S. and T. Lindsey (2006) 'Intellectual property, civil law and the failure of law in Indonesia: Can criminal enforcement of economic law work in developing countries?' in T. Lindsey (ed.) *Law Reform in Transitional and Developing States*, London: Routledge.

Butt, S. and T. Lindsey (2012) *The Indonesian Constitution: A contextual analysis*, Oxford: Hart.

Cassese, A. (1986) *International law in a divided world*, Oxford: Oxford University Press.

Colbran, N. (2010) 'Realities and challenges in realising freedom of religion or belief in Indonesia', *International Journal of Human Rights* 14: 678.

Eldridge, P. J. (2002) 'Human rights in post-Soeharto Indonesia', *Brown Journal of World Affairs* 9(1): 127.

Hoof, F. V. (1996) 'The impact of international law in the legal order of the Netherlands: The role of the judiciary' in C. Saunders (ed.) *Courts of final jurisdiction: The Mason court in Australia*, Annandale: Federation Press.

Jetschke, A. (2011) *Human rights and state security – Indonesia and the Philippines*, Philadelphia: University of Pennsylvania Press.

Juwana, H. (2005) 'Treaty making – Indonesian practice seminar on the parliamentary oversight of treaties, Republic of Indonesia & Australia', Jakarta, 15–16 July.

Juwana, H. (2010) *Hukum internasional dalam perspektif Indonesia sebagai negara berkembang*, Yarsif Watampone.

Kartasapoetra, G. and R. G. Kartaputra (1984) *Indonesia dalam lingkaran hukum internasional (dari abad ke abad)*, 1st edn, Sumur Bandung.

Kusumaatmadja, M. (2003) *Pengantar hukum internasional*, 2nd edn, PT Alumni.

Manan, B. (1985) 'Kekuasaan presiden dalam masalah dan hubungan internasional', *Majalah Padjadjaran* (No. 1–2).

Parlina, I. (2013) 'Minorities wary of mixed signals from Court', *Jakarta Post*, 3 May.

Roestandi, A. (2004) 'Mengapa saya mengajukan Dissenting Opinion' in R. Harun, Z. A. M. Husein and Bisariyadi (eds) *Menjaga Denyut Konstitusi: Reflexi satu tahun Mahkamah Konstitusi*, Konstitusi Press.

Saragih, B. B. T. and M. S. Aritong (2013) 'RI to ratify Rome Statute', *Jakarta Post*, 6 March.

Schermers, H. (1987) 'Netherlands' in F. Jacobs and S. Roberts (eds) *The effect of treaties in domestic law*, London: Sweet & Maxwell.

Sefriani (2011) *Hukum internasional: suatu pengantar*, 2nd edn, Rajawali Press.

Sidik, S. (2004) *Hukum internasional dan berbagai permasalahannya (suatu kumpulan karangan)*, Lembaga Pengkajian Hukum Internasional Fakultas Hukum Universitas Indonesia.

Sunaryati, H. (2009) *Analisis dan evaluasi hukum tentang ratifikasi perjanjian internasional di bidang hak asasi manusia dan urgensinya bagi Indonesia*, Departemen Hukum dan Hak Asasi Manusia.

Triggs, G. D. (2010) *International law: Contemporary principles and practices*, 2nd edn, Chatswood: LexisNexis.

UN Human Rights Council (2012) 'Report of the Working Group on the Universal Periodic Review: Indonesia', UN Doc A/HRC/21/7, 5 July.

Wayan, P. (1987) *Beberapa maslaah dalam hukum internasional dan hukum nasional Indonesia*, 1st edn, Binacipta.

Wibisana, A. G. (2011) 'The development of the precautionary principle in international and in Indonesian environmental law', *Asia Pacific Journal of Environmental Law* 14: www.ssrn.org.

Zhang, D. (2010) 'The use and misuse of foreign materials by the Indonesian Constitutional Court: A study of Constitutional Court Decisions 2003–2008', University of Melbourne, January.

6 National regulatory framework for REDD+ in Indonesia

In this chapter, we provide an overview of the key national legal instruments that directly regulate or otherwise affect REDD+ activities in Indonesia. We begin by outlining the national 1999 Forestry Law – the statute from which most subsequent REDD+-related regulatory instruments draw their legal legitimacy – before assessing REDD+-related laws issued by the President and the Forestry Ministry.[1]

The Forestry Law (1999)

Law 41 of 1999 on Forestry, as amended by Law 19 of 2004 (the Forestry Law), is the primary legal instrument under which forest lands are managed, exploited and protected in Indonesia. It was enacted by Indonesia's national parliament (*Dewan Perwakilan Raykat*, or DPR) – a body of 550 elected representatives.[2]

The primary starting point of the Forestry Law is that, by virtue of Article 33(3) of the Constitution, the State controls forest areas (*kawasan hutan*). Article 33(3) of the Constitution states:

> The earth and water and the natural resources contained within them are to be controlled by the state and used for the greatest possible prosperity of the people.

Article 4(1) of the Forestry Law echoes this provision, declaring that:

> All forest in the territory of the Republic of Indonesia, including the natural resources contained within them, are controlled by the state for the greatest possible prosperity of the people.

---

1 We encourage reading this chapter in conjunction with Chapter 7 on jurisdictional conflicts, which discusses the relative status of the types of law used in the Indonesian legal system.
2 In addition to enacting legislation, the national parliament reviews the actions of the executive and holds the President and Ministers accountable for their actions and policies. Once statutes are approved by the national parliament, the President signs them into law. Recent constitutional amendments prevent the President from refusing to endorse a DPR-approved bill. The amendments provide that if the President does not sign a bill approved by the DPR within 30 days, the bill is automatically passed into law.

Articles 4(2)(a) and 4(2)(c) give the State broad authority to 'regulate and administer everything relating to forests, forest areas, and forest products' and to 'regulate and determine legal relationships between people and forests, and regulate legal activities concerning forestry'. The Law also grants the State power to stipulate particular territory as forest area and even to classify 'forest areas as non-forest areas' (Article 4(2)(b)). When exercising this authority, however, the State is to observe the rights of pre-existing traditional communities in the forest areas, provided that observing those rights does not contradict the national interest (Article 4(3)).

The Law divides Indonesian forest areas into two types: forests that have rights or concessions awarded over them (*hak hutan*) (Article 1(5)) and the remainder, which are classified as State forest (*hutan negara*) (Article 1(4)). However, we note that Indonesia's Constitutional Court has, in a 2013 decision, effectively amended this provision to recognise so-called 'customary forest' (*hutan adat*) as a separate forest category, falling within what the Forestry Law had previously classified as State forest. We discuss this in detail in Chapter 9.

The Law also establishes three categories of forest, distinguished by the functions or purposes for which they can be used under the Forestry Law and its various implementing regulations, discussed later: production forest, protected forest and conservation forest.

Article 1(8) defines 'production forest' as 'forest area with the primary function of producing forest products'. In production forests, 'environmental services', timber and non-timber exploitation and timber and non-timber resource harvesting can take place. Production forests are, therefore, the main forest areas used for industrial production of timber and agricultural products such as palm oil. Those seeking to engage in these activities require a permit (Article 28(2)).[3]

'Protected forests' have the primary function of 'protecting life support systems, regulating water flow, preventing floods, controlling erosion, preventing sea water intrusion and protecting soil fertility' (Article 1(8)). Environmental protection services can be provided in these forests, and non-timber forest resources can be harvested from them (Article 26(1)). However, the Forestry Law does not allow commercial logging in protected forests. Again, a permit is required to perform these activities in protected forests (Article 26(2)).

Conservation forests are defined in Article 1(9) as forest areas with 'special characteristics', having the main function of preserving plant and animal

---

3 The specific permits mentioned are: permission to exploit forest areas (*izin usaha pemanfaatan kawasan*), permission to benefit from environmental services (*izin usaha pemanfaatan jasa lingkungan*), permission to exploit timber products (*izin usaha pemanfaatan hasil hutan kayu*), permission to exploit non-timber products (*izin usaha pemanfaatan hasil hutan bukan kayu*), permission to harvest timber products (*izin pemungutan hasil hutan kayu*) and permission to harvest non-timber products (*izin pemungutan hasil hutan bukan kayu*).

diversity as well as ecosystems. Conservation forests that also contain life support systems are classified as 'nature reserves' (Article 1(10)); and nature reserves that also have non-depleting natural biological resources can be classified as 'nature conservation forests' (Article 1(11)). Hunting reserves are also categorised as 'conservation forests' (see Article 7).

The Forestry Law grants strong forest-related powers to the State, which the Forestry Ministry primarily claims and exercises. These include powers to designate areas as forest (Article 1(3)) and to classify forest areas as production, protection or conservation forests (Article 6(2)). (However, as we shall see in Chapter 9, the Constitutional Court has effectively amended Article 1(3), deciding that 'designation' is simply one step in a more complex process that requires the Ministry to work with local governments.) The Forestry Ministry can also change forest 'allocations and functions' if deemed necessary to 'meet the demands of national development … and societal aspirations'.[4] (However, there is very little incentive for the Forestry Ministry to declassify areas as 'forest' because by doing so, the Ministry relinquishes its authority to award concessions or licences over those areas.[5]) The Law also authorises the Forestry Minister to permit mining in forest areas by issuing lease-use licences (Article 38(3)), but only in production or protection forests (Article 38(1)).[6]

The Forestry Law prohibits various activities, including:

- causing damage to the forest, even by concession or licence holders (Articles 50(1) and 50(2))
- illegally working, using or occupying forest territory (Article 50(2)(a))
- clearing forest areas (Article 50(2)(b))

---

4 Government Regulation 10 of 2010 on Procedures to Change the Allocation and Function of Forest Areas grants this power to the Ministry. This Regulation defines changing allocations as changing an area classified as a forest area into a non-forest area (Article 1(13)). Changing functions are defined as changing part of or an entire forest from one category to another (Article 1(14)), although the Regulation requires that this be preceded by research conducted by a competent government authority, with involvement from 'related parties' (Articles 1(19), 2 and 5; Indrarto et al. 2012: 37). Under Article 43(2) of this Regulation, changes must be 'based on recommendations and submissions' from the head of the local government in which the change is proposed.

5 Declassification of forest areas allows local government to issue licences for crops and mining under, respectively, Law 18 of 2004 on Plantations and Law 4 of 2009 on Mineral and Coal Mining (Indrarto et al. 2012: 31). Law 26 of 2007 on Spatial Planning also allows regional governments to allocate land for non-forestry purpose areas that have not been allocated as forest. They are permitted to issue licences in those areas for productive activities, such as plantations (Wells et al. 2012: 12).

6 Government Regulation 24 of 2010 on Use of Forest Areas fleshes out Article 38(3)). The Regulation declares that forest areas can be used for non-forestry related activities (Article 2), but only for 'unavoidable' strategic purposes (Article 4(1)), one of which – according to the Regulation – is mining (Article 4(2)(b)), including for oil and natural gas, minerals, coal and geothermal resources (Elucidation to Article 4(2)(b)). Other purposes include religion; electricity, telecommunications and transport infrastructure; public facilities; defence and security; and 'forestry-related industry' (Article 4(2)).

- logging in close proximity to various water sources (Article 50(2)(c))
- felling trees or harvesting forest produce without having a concession or permit from a relevant government official (Article 50(2)(e))
- receiving, buying, selling, trading, storing or possessing forest produce that is known or ought to be known to have been illegally obtained or harvested from forest areas (Article 50(2)(f)).

Article 78 imposes significant terms of imprisonment and fines for breach of these prohibitions. For example, those who fell trees or harvest produce without a concession permit (thereby breaching Article 50(3)(e)) face a maximum of 10 years' imprisonment and a maximum fine of Rp 5 billion.

REDD-related laws issued by the President of Indonesia

The President is the Indonesian head of State and head of government. He or she is directly elected as a 'pair' with the Vice President for a maximum of two five-year terms. The President can issue Presidential Regulations (*peraturan presiden*) usually, but not exclusively, to implement statutes enacted by the national parliament. The President's signature is also required for government regulations (*peraturan pemerintah*) to come into force. The President leads the National Council on Climate Change, which was established in 2008 under Presidential Regulation 46 of 2008. The Council's main function is to develop policies, strategies and programmes to address climate change and to coordinate climate change activities.

The President at the time of writing, Susilo Bambang Yudhoyono, has strongly supported environmental and forestry reforms, including REDD+, and has issued the Presidential laws summarised in this section. He has pledged that Indonesia will reduce greenhouse gas emissions by 26 per cent by 2020 and by 41 per cent with international assistance. He also signed a moratorium on issuing new concessions with Norway. However, Yudhoyono's second five-year term was to expire in October 2014. The future trajectory of REDD+ in Indonesia largely depends on his successor's commitment to the REDD+ concept. It will also hinge on his or her ability to handle the Forestry Ministry (discussed later) and to resist the short-term lure of revenue from competing industries, particularly the palm oil sector.

REDD+ Taskforce Decree (2010)

The REDD+ Taskforce was established by Presidential Decree 19 of 2010 as a temporary body that would ultimately be replaced by a REDD+ Agency. At time of writing, this Agency had not yet been established and, in the interim, the Taskforce's term had already been twice extended.[7]

---

7 By Presidential Regulation 25 of 2011 and Presidential Decree 5 of 2013, amending Presidential Regulation 25 of 2011.

Functioning under the Presidential Working Unit for Development Supervision and Management (*Unit Kerja Presiden untuk Pengawasan dan Pengendalian Pembangunan*, or UPK4), the Taskforce has many functions. These include coordinating Indonesia's active and pilot REDD+ projects, overseeing the Norway moratorium, deciding on REDD+ funding mechanisms and establishing the REDD+ Agency. The Taskforce is also responsible for coordinating the National Action Plan to Reduce Greenhouse Gas Emissions by 2020 (Articles 3 and 4).

National Greenhouse Gas Emissions Reduction Regulation (2011)

Presidential Regulation 61 of 2011 on the Action Plan to Reduce National Greenhouse Gas Emissions (RAN-GRK)[8] is directed at ministries, non-Ministerial institutions and regional governments (Article 3), but also functions as a 'guide' for the public and commercial enterprises (Article 4).

The Schedule to the Regulation directs various entities to perform particular acts in specified provinces, to help reduce emissions to meet the 26 and 41 per cent targets. As for the forestry and peatland sectors, for example, the Schedule directs the Forestry, Public Works and Agriculture Ministers to – in most provinces of Indonesia – clearly demarcate and designate forest areas, improve peatland and forest reclamation and rehabilitation, employ sustainable agricultural practices, prevent forest fires, handle cases of illegal logging and mining, protect conservation areas in protected forests and 'enhance plantation forest businesses'. It also sets emission reduction targets in five sectors: forestry and peatland, agriculture, energy and transportation, industry and waste management. These activities are to be funded by national and regional governments (Article 11), although the Regulation does not allocate financial responsibilities between them.

Greenhouse Gas Inventory Regulation (2011)

Articles 63(1), 63(2) and 63(3) of Indonesia's primary environmental protection statute (Law 32 of 2009 on Environmental Protection and Management) require the central, provincial and district/city governments to produce inventories of their natural resources and greenhouse gases. Presidential Regulation 71 of 2011 on the Implementation of a National Greenhouse Gas Inventory seeks to 'implement' these provisions.[9] This Regulation provides for the periodic monitoring of emissions and carbon stock volumes at national, provincial and district/city levels (Articles 2, 9 and 10).[10]

---

8 A translation of this President Regulation is available at: http://forestclimatecenter.org/files/2011-09-20%20Presidential%20Regulation%20No%2061%20on%20The%20National%20Action%20Plan%20for%20Greenhouse%20Gas%20Emission%20Reduction.pdf.

9 Preamble, Part C of Presidential Regulation 71 of 2011.

10 This Regulation appears to have several shortcomings. For example, it does not specify the regularity with which such tests and calculations must be performed. Also, important matters

Moratorium Continuation Instruction (2013)[11]

This Presidential Instruction continues the Norway moratorium, against the issuance of new forest concessions over natural forest or peatlands within areas of conservation forest, protection forest and some types of production forest (Part I). The Instruction extends the moratorium until 13 May 2015 (Article 5).

Those who already hold concessions over areas that fall within the moratorium areas can continue their exploitation and other activities and can even seek to extend these concessions (Part II). Also exempt from the operation of the moratorium are areas where vital national development or ecosystem restoration is required (Part II). The Instruction orders the Presidential Working Unit and the REDD+ Taskforce, or another REDD+-specific agency, to monitor the moratorium and report to the President every six months (Articles 6–7). The Instruction also directs the Forestry Ministry to revise the indicative map for the areas covered by the moratorium (Article 3(1)).

REDD+-related laws issued by the Forestry Ministry

The Forestry Ministry is the national institution primarily responsible for managing Indonesia's forest areas. Originally a division within the Agriculture Ministry, it was established as a separate department in 1983.[12] The Ministry has issued many Regulations relevant to REDD+ activities and works in conjunction with the REDD+ Taskforce and project groups.

REDD+ Demonstration Activities Regulation (2008)[13]

In this brief Regulation, the Forestry Ministry provided a legal basis for the establishment of REDD+ demonstration activities in Indonesia. Under the Regulation, demonstration activities can be performed in State forests, or forests over which concessions or licences have been granted (Article 3). They can be run by the government, holders of forest timber licences, holders and

are left to future Ministerial regulation, such as verification of emissions (Article 6(2)) and the obligation imposed on commercial enterprises to disclose their inventory data to the relevant local government (Articles 15(2) and 18). This is significant, because the Regulation does not specify which ministry should issue these Ministerial regulations. Article 1(16) defines 'Minister' as the 'Minister who manages government affairs in the field of environmental protection and management'. This could encompass not only the Environment Ministry, but also the Forestry, Public Works and Agriculture Ministers, as well as others whose portfolios affect the environment. The likely result is that ministries will either issue inconsistent verification regulations, or will justify the failure to issue such regulations on the basis that issuing them is the responsibility of another ministry.

11 Presidential Instruction 6 of 2013 on the Moratorium on Forest Concessions and Improved Management of Primary Natural Forest and Peatlands.

12 See Presidential Regulation 15 of 1984.

13 Forestry Minister Regulation 68 of 2008 on the Establishment of Demonstration Activities for Reducing Carbon Emissions from Deforestation and Forest Degradation.

managers of 'rights' forests and managers of customary forests (Articles 1(6) and 4(1)). Under the Regulation, the main purposes of demonstration activities are testing and developing sustainable forestry methodologies, technologies and institutions as part of efforts to reduce carbon emissions by avoiding deforestation and forest degradation (Article 2(1)).

The Regulation outlines the procedures for applying to run these demonstration activities and for the assessment of applications by the Forestry Ministry's Climate Change Working Group (*Kelompok Kerja Pengendalian Perubahan Iklim*, or POKJA PI).[14] Article 5(3) of the Regulation states that demonstration activity criteria and feasibility indicators will be further specified by a future Ministerial regulation. This regulation appears to be the 2012 Forestry Minister Regulation on Forest Carbon Management, discussed later.

If the Forestry Minister approves an application to run a demonstration activity, he or she can stipulate the location and scope of the activity, duration of the programme (although this may not exceed five years) and conditions relating to risk mitigation and the distribution of proceeds (Article 5(6)).

REDD Mechanisms Regulation (2009)[15]

This instrument represents the first regulatory attempt to provide a basic legal framework for REDD+ activities in Indonesia. It specifies that REDD+ projects can be conducted in:

- natural, plantation, community and community plantation forests over which timber extraction concessions have been awarded
- ecosystem restoration timber forest production areas
- production forest management areas
- conservation, customary, private and village forests (Article 3).

REDD+ projects can run for up to 30 years and can be renewed (Article 13).

The Regulation distinguishes between international and domestic REDD 'participants' (Article 4(1)). National concession holders and forest managers can apply to participate in REDD activities, as can international governments, commercial enterprises and funding organisations (Articles 4(2) and 4(3)). To participate, concession holders must be able to produce a copy of the concession, a recommendation from the regional government with territorial jurisdiction over the proposed site, have a REDD+ implementation plan and fulfil various 'location criteria' (Articles 5–9). By contrast, a traditional community wishing to participate in REDD+ activities must obtain a formal Decree from the Forestry Ministry acknowledging the existence of that community.

---

14 See Article 5 of the Regulation. Applications must include area location maps stating the type, duration and value of activities, as well as detailing risk management and revenue allocation.
15 Minister of Forestry Regulation 30 of 2009 on Mechanisms for Reducing Emissions from Deforestation and Forest Degradation.

Once the Forestry Minister receives an application, he or she is to submit it to the 'REDD+ Commission' to review the feasibility of the proposed activity. The Regulation sets out the criteria by which this Commission must evaluate the activity. These criteria include the availability of relevant data and information; biophysics and ecology, including carbon stocks and biodiversity; the threat to forest resources; cultural, social and economic considerations; economic feasibility; and management – including efficiency, effectiveness and the applicant's commitment to REDD+.[16]

Certificates evidencing emission reductions from REDD+ activities can be awarded, provided that the emissions have been independently verified (Articles 16 and 17). National entities can then receive payments from international entities for emissions reductions (Article 14(1)) and international entities can purchase certificates to fulfil their emissions reduction commitments (Article 14(2)).[17]

Carbon Sequestration and Storage Regulation (2009, amended 2013)[18]

This Regulation governs carbon sequestration (*RAP Karbon*) and storage (*PAN Karbon*), as 'environmental services'. Under Article 3, these activities include postponing logging, maintenance, rehabilitation, planting, expanding conservation areas and applying harvest rotation. Indrarto et al. (2012: 34) provide a useful summary of entities entitled to apply to provide these services – those who:

1   hold forest timber concessions for natural, plantation or community plantation forests
2   hold use permits for protection forests or community forests
3   are village forest managers.

In areas not subject to permits, individuals, cooperatives and other businesses operating in agriculture, estate crops or forestry may also submit proposals for carbon capture and/or sequestration enterprises (Article 7).

The Regulation also stipulates prerequisites and procedures for submitting applications and obtaining approvals for activities in particular areas (Articles 5–8). Ministerial permission will usually be required to provide these environmental services over land already subject to a concession (Article 6). However, authorisation can be obtained from governors, district heads

---

16   *Ibid.*, Sch. 2
17   The Regulation contains six Schedules. These are guidelines for regional governments on providing recommendations (Sch. 1); selecting REDD locations (Sch. 2); preparing REDD implementation plans (Sch. 3); evaluating REDD applications (Sch. 4); determining emissions references, monitoring and reporting (Sch. 5); and verifying REDD activities (Sch. 6) (Indrarto et al. 2012: 76).
18   Forestry Ministry Regulation 36 of 2009 on Permit Procedures for Carbon Sequestration and/or Storage Enterprises in Production Forests and Protection Forests.

Table 6.1 Distribution of proceeds from carbon sequestration and storage

Licence holder/developer	Distribution		
	Government	Community	Developer
Timber exploitation (natural and plantation forests)	20%	20%	60%
Ecosystem restoration (natural forest)	20%	20%	60%
Timber exploitation (community plantation forest)	20%	50%	30%
Community forest	20%	50%	30%
Customary forest	10%	70%	20%
Village forest	20%	50%	30%
Forest management unit	30%	20%	50%
Special Purpose Forest Area	50%	20%	30%
Protected forest	50%	20%	30%

Source: Forestry Minister Regulation 36 of 2009, Schedule III

or mayors to perform activities on land not subject to a concession, such as industrial plantation forests (*hutan tanaman industri*), ecosystem restoration areas and community plantations (Articles 9–11).

The Regulation anticipates that these services can form part of REDD schemes, but requires further regulation by the Minister (Article 4). Activities are to be verified by an independent verification institution and, if this institution confirms the emissions sequestered or stored, certificates can be issued for trade on carbon markets (see Articles 14–17).[19] Article 15(5) states, however, that the procedures for transferring and granting carbon credits are to be detailed in later Ministerial regulations.

Schedule II to the Regulation outlines Standards on Project Development and Carbon Markets. Included are various approved methodologies for assessing carbon levels, price ranges and independent verification methods.

Schedule III to the Regulation sets out how proceeds from carbon sequestration and storage are to be distributed between the State, affected community and project developer (see Table 6.1). For projects performed on land subject to most types of concessions, the developer will receive 60 per cent of the proceeds, with the government and community receiving an equal share of the remainder. However, for traditional and village forests, the community obtains a greater proportion – between 50 and 70 per cent, depending on the type of forest.

Schedule III also specifies that the 'government' share is to be divided as follows: the central government receives 40 per cent, provincial governments 20 per cent and district governments 40 per cent.

Forestry Ministry Regulation 11 of 2013 amended this 2009 Regulation. It provides that verifications are to be registered with the National Registration Board or on the Voluntary Carbon Market to obtain a verified emission

19 Article 14(7) allows Developer Projects to be registered with the National Land Board or on the international Voluntary Carbon Market (VCM) to obtain a verified emission reduction (VER).

reduction certificate (Article 15(1)). The amendments also specify that the Forestry Ministry Secretary-General is to register verifications for carbon storing or sequestration until the Board is established (Article 15(2)). VER certificates can be sold directly by project developers or on the carbon exchange market, either in Indonesia or internationally (Article 15(3)).

Regulation on Devolution of Forestry Governance to Governors (2013)[20]

This Regulation devolves some forestry-related tasks from the Forestry Ministry to governors of Indonesia's 33 provinces (Article 3(4)). The Schedule to this Regulation lists the specific functions to be allocated to each governor by 31 December 2013. These include development, supervision and control of:

- management of plantation forests and natural forests
- management planning for production forests
- distribution of forest products and forestry contributions
- primary forest commercial industries
- forest and land rehabilitation (including mangroves, swamp, peat moss and beaches) and rehabilitation agencies.

The Schedule also seeks to gradually devolve:

- implementation of integrated river networks
- establishment of working areas and concessions over community and village forests
- supervision and control of community and village forests
- supervision, facilitation, monitoring and evaluation of forestry education
- raising awareness about forest area boundaries
- campaigns to eradicate illegal logging and trade in wildlife and fauna
- increasing human resources capacity to protect forests, including forestry police and civil servant investigators.

The Forestry Ministry is to provide funds to governors to perform these devolved functions (Article 4(5)) and governors are required to establish provincial working groups to perform them (Article 4(1)).

The Regulation allows the Ministry to resume authority over the matters devolved if policies change or governors fail to observe Ministry standards and procedures (Article 6(5)). The Regulation also prohibits governors from further delegating their tasks to regents, mayors or village heads (Article 3(3)).

This Regulation appears likely to significantly affect the management of forests designated for REDD because it will presumably place responsibility

20 Forestry Ministry Regulation 25 of 2013 on Devolution of Forestry Governance in 2013 to 33 Provincial Governors as Government Representatives.

on governors to 'manage' them. However, at time of writing, no information was available about how this Regulation was being implemented, if at all.[21]

Forest Carbon Management Regulation (2012)[22]

At time of writing, this was the most recent Forestry Minister Regulation dealing with demonstration activities (Article 1(2)) and 'forest carbon activities' (Article 1(3)). Significantly, this Regulation grants the Forestry Ministry power to issue a new form of concession: the Forest Carbon Management Concession (FCMC, or *Izin Penyelenggaraan Karbon Hutan*) (Article 1(8)). Article 12(4)(b) takes the role of evaluating demonstration activities from the Climate Change Working Group, which the REDD Demonstration Activities Regulation (2008), mentioned above, had allocated to it. Forest carbon activities are defined as activities for the storage or sequestration of carbon, including:

- cultivation, planting, forest management and sustainable harvesting
- extending felling cycles and enriching plantings in timber concession areas
- protecting and safeguarding timber concession areas
- biodiversity preservation
- managing protected and conservation forests (Article 3(2)).

These activities can be performed in customary and concession forests and in State production, protection or conservation forests (Article 3(3)).

Demonstration activities must develop processes to create or improve measurement, implementation and reporting standards, or must support the development of those standards (Article 4(1)). Applications for permission to perform these activities must be submitted to the Forestry Minister. Applications must include the proposed location, the form and duration of cooperation, estimated value of the activity and risk management considerations (Article 4(2)). If project managers wish to engage in 'forest carbon activities' as part of these demonstration activities, they must obtain an FCMC from the Minister.

The Regulation allows FCMCs to co-exist with other types of forest concession and specifies that holders of those concessions can initiate and implement forest carbon activities. For forest areas that are concession free but have an identified 'manager' – such as in production, protection, conservation, traditional or people's forests – an FCMC can still be obtained for carbon storage and sequestration (Articles 6 and 7).[23]

---

21 However, the Ministry had issued guidelines on the devolution, e.g. Forestry Ministry Regulation 27 of 2013 on Technical Guidance on the Implementation of the 2013 Devolution of Forestry Affairs to Governors as Representatives of the National Government.
22 Forestry Minister Regulation 20 of 2012 on Management of Forest Carbon.
23 The Climate Change Working Group is to oversee the implementation of these FCMC licences (Article 6(3)).

To maintain carbon stocks, FCMC concession holders are required to protect the forest areas over which they hold the concession from damage, fire, clearing and over-cultivation. They must also sustainably manage the forest (Article 8(6)). However, the Regulation does not prescribe sanctions for non-compliance.

The Regulation also permits FCMC holders to 'trade' carbon on the international and voluntary markets (Article 9(1)(b)). This means they can trade the amount of carbon 'saved' in a particular year, assessed by reference to a baseline (Article 8(3)). However, this Regulation leaves details about carbon trading mechanisms to yet another future Ministerial regulation (Article 8(4)). To our knowledge, this regulation had not yet been issued.

The Regulation is silent on how funds from carbon trading should be distributed.[24] Presumably, then, the distributions listed in Schedule III to the Carbon Sequestration and Storage Regulation (2009), mentioned earlier, continue to apply, even though their validity is challenged, particularly by the Finance Ministry.

References

Indrarto, G. B., P. Murharjanti, J. Khatarina, I. Pulungan, F. Ivalerina, Justitia et al. (2012) *The context of REDD+ in Indonesia: Drivers, agents and institutions*, Bogor, Indonesia: CIFOR and ICEL.

Wells, P., N. Franklin, P. Gunarso, G. Paoli, T. Mafira, D. R. Kusumo et al. (2012) 'Indonesian Constitutional Court Ruling Number 45/PUU-IX/2011 in relation to forest lands', *Daemeter/TBI Indonesia/Makarim & Taira S.*

24 Government income from carbon trading will be deemed 'non-tax State income' (*pendapatan negara bukan pajak*).

7 Jurisdictional conflicts and REDD+[1]

The restructuring of the Indonesian polity from 1998 resulted in power being dispersed among many institutions of government, both national and regional. This fragmentation has led to immeasurable complexities in many areas of Indonesian law, particularly when different institutions and tiers of government claim authority to regulate the same subject matter. In the face of inconsistent laws, which law must legal subjects follow? Can one of the inconsistent laws be invalidated in the event of any inconsistency? Unfortunately, these questions, although fundamental in any modern State, are not easy to answer in Indonesia, and have major implications for the design and functioning of any REDD+ scheme.

Disputes about the jurisdictional reach of various government institutions arise commonly in many States, both developing and developed. They are particularly common in federal and unitary States in which power has been de-concentrated or decentralised. Many States have an independent body with the power to decide jurisdictional disputes by reference to specified rules and processes. Without the involvement of an impartial adjudicator there will be constant claims that each institution interferes in the affairs of the other. In the Indonesian context, this is manifested in the refrain from many regional governments that the central government perpetuates the Soeharto era practice of exerting excessive control over the regions. The lack of a mechanism to demarcate jurisdictional authority and resolve jurisdictional disputes leaves citizens in a precarious position: needing to choose which law to follow and which to ignore. Some of these types of dispute have played out in Indonesia's Constitutional Court (see further the discussion in Chapter 9). In this chapter, we focus on disputes that fall outside the Constitutional Court's jurisdiction that have relevance for the implementation of REDD+. Although we discuss the resolution of jurisdictional disputes between national-level institutions, our main concern is disputes arising between tiers of government, whether national against regional or regional against regional. As we shall see, in Indonesia two mechanisms exist for some of these disputes – bureaucratic review by the central government for disputes between the

1 Draws on Butt (2010).

central and regional governments and judicial review by the national Supreme Court (*Mahkamah Agung*, or MA) for those and other national-level disputes. Both are deeply flawed and of limited utility in most instances. Worse, for most types of disputes – including disputes between national ministries – there are no mechanisms for resolving conflicts and inconsistencies.

Disputes over natural resources, including forests, are particularly fraught, with various government institutions (both national and regional) actively seeking to regulate and administer those resources to secure proceeds from their exploitation. Much is at stake, because jurisdiction to control and permit exploitation and extraction brings income for government institutions, which may be necessary to supplement their operational incomes – without which they might be unable to operate effectively, if at all. Such power also increases the potential for officials to extract bribes.

National-regional jurisdictional conflict

In 1999, the year following President Soeharto's resignation, Indonesia's national parliament enacted the country's first statute on regional autonomy. Under these reforms, the central government devolved authority to local government institutions to govern their own territorial jurisdictions, including by making their own policies and laws. Four 'levels' of regional governments received powers under the reforms: provinces (*propinsi*), districts (*kabupaten*), cities (*kota*) and villages. (Districts and cities are the same 'level' of government, distinguished primarily by their population density and the size of their territory.) Only several matters, discussed later, remained within the exclusive jurisdiction of the central government.

Provinces, districts and cities have their own legislatures and executives, both with lawmaking powers. The laws they produce are popularly described as *perda*. This is an abbreviation of *peraturan daerah*, literally, 'regional regulation', but often translated as 'by-law'. We note, however, that in legal terms, *perda* refers only to laws passed by local legislatures, whether provincial, district or city. A reference to a provincial *perda*, for instance, is a reference to a law enacted by a provincial parliament. By contrast, laws passed by the head of the executive arm of government – governors (in provinces), regents (in districts) and mayors (in cities) – are generally referred to as regulations of heads of regions (*peraturan kepala daerah*) or more specifically as *peraturan gubernur* (governor regulation), *peraturan bupati* (regent regulation) and *peraturan walikota* (mayoral regulation).

One result of regional autonomy was the proliferation of lawmaking bodies, the numbers of which have continued to increase. In 1998 – the year before the first batch of decentralisation laws was passed – Indonesia had approximately 292 local governments outside Jakarta (Fitriani et al. 2005: 58).[2] In subsequent years, the number has grown as provinces, municipalities

2 Smaller government units, such as villages, were granted regulatory powers, albeit limited.

and cities split into two or more (Booth 2011). In 2003 – the year before a second batch of decentralisation laws were enacted – this number had grown to around 440 municipalities and districts (Fitriani et al. 2005: 57; Smoke 2005). By 2012 there were 34 provinces as well as 500 municipalities and districts.[3] As of early 2014 well over 1000 Indonesian institutions and individuals had powers to issue binding laws – at least 50 per cent more than when regional autonomy was designed and put into place. This number is likely to increase further as provinces, municipalities, districts and other tiers of government are carved out from existing regions.

Law 32 of 2004 on Regional Government (the 2004 Law) provides local governments with jurisdiction to make laws on a wide variety of issues and requires them to exercise it. Adopting terminology employed in Article 18(5) of the Constitution, Article 2(3) of the 2004 Law gives local governments the 'broadest autonomy' in order to 'increase the prosperity of the community, public services and regional competitiveness'. The Law also provides provincial, district and city governments with various 'obligations' (*urusan wajib*), including development planning, public order and security, infrastructure, health, education, labour, small and medium enterprises, the environment, land affairs, public administration and investment (Articles 13(1) and 14(1) of the 2004 Law).

The 2004 Law imposes two limits on regional government lawmaking. First, as mentioned, the Law reserves some matters exclusively for the central government, including foreign affairs, defence, security, national monetary and fiscal matters and religion (Article 10(3)). The central government also has exclusive power over justice-sector matters (*urusan yustisi*), including establishing judicial institutions, appointing judges and prosecutors, determining justice-sector and immigration policy, and enacting statutes and other national laws (Elucidation to Article 10(3)). The central government can, however, delegate its jurisdiction to regulate these matters to local governments (Article 10(4)). The second limitation on regional lawmakers is that the central government has retained power to legislate in the areas that Article 10(3) of the 2004 Law does not mention (Bell 2001: 29). In other words, the jurisdiction granted to local governments is not exclusive. Under Article 10(5), the central government can itself exercise the jurisdiction granted to local governments or delegate this to regional apparatuses and officials. Legally speaking, the central government can continue to regulate any matter over which regional governments also have jurisdiction. Significantly, many types of national government law will trump local laws, rendering them invalid to the extent of any inconsistency. We return to discuss the implications of Article 10(5) and the 'hierarchy of laws' later in this chapter.

As for jurisdiction over forestry, Government Regulation 62 of 1998 on Delegation of Forestry Governance to Regional Governments gave authority to district governments to 'carry out rehabilitation and reforestation

3 See the Ministry of Home Affairs Website: www.depdagri.go.id.

activities, soil and water conservation, protected forest management, extensions and small-scale community forest activities' (Indrarto et al. 2012: 27). However, the 2004 Autonomy Law is largely silent on forestry jurisdiction and the 1999 Forestry Law seems to contradict the 1998 Regulation. Article 48(2) of the Forestry Law states that: 'Forest protection in state forests is to be implemented by the government' and Article 1(14) defines 'government' as the 'central government'.

Yet Government Regulation 45 of 2004 on Forest Protection states that 'forestry protection falls within the jurisdiction of the Central government and/or Regional Governments' (Article 3(1)). This is significant because the Regulation also sets out various obligations of the central government, regional governments and communities, including various 'principles of forest protection' to which adherence appears to be required (Article 6). These principles include preventing or limiting damage to forests and upholding forest rights (Article 6), presumably including those of traditional communities.[4]

Providing the same statutory tasks to three tiers of government and the community, without further specific allocations of these tasks among them, appears to have led to many jurisdictional overlaps and conflicts. Below, we outline some of the difficulties presented in resolving these.

We now turn to discuss how some local governments have used their lawmaking powers in respect of forestry matters, sometimes in apparent conflict with national law.

Regional logging licences

Some commentators have observed that, particularly in the early days of decentralisation, local government officials in forest-rich regional areas used their new lawmaking powers to allow themselves to authorise logging and other forest activities. According to Barr et al.:

> The enthusiastic efforts of *kabupaten* [district] governments to establish direct administrative control over the forests within their jurisdictions generated a plethora of district regulations, or *perda*, which often stood in stark opposition to the national government's forestry laws. In some cases, the new district regulations provided varying degrees of legitimacy to timber that was harvested without permits from government agencies at any level. Operationally, many districts also showed that they had little capacity for regulating the activities of the timber companies that received district logging and conversion permits. This led critics,

4 They must also perform various tasks, including: to increase community productivity, facilitate the establishment of community institutions, increase the role and participation of the community in forest management, cooperate with concession or licence holders, increase coordination of forest protection activities, help create alternative livelihoods for communities, increase the effectiveness of reporting 'security disturbances' in forests and impose sanctions for breaches of the law.

particularly in [the Forestry Ministry], to charge that district governments had effectively used the decentralization process to 'legalize' illegal logging. In many cases, as well, district governments allocated logging licenses and forest conversion permits in sites that directly overlapped with areas previously assigned to HPH-holders (2006: 88) … many of the *perda* issued by district governments to regulate timber production within their jurisdictions directly contradict regulations issued by the MoF [Ministry of Forestry], thereby appearing to authorize practices that the central government considers to be illegal. The fact that there has not been a clear mechanism for resolving such legal-regulatory contradictions in a timely manner has been a major impediment to Indonesia's decentralization process. (Barr et al. 2006: 99)

To be fair, some of the 'early' *perda* in which regional governments sought to grant themselves power to issue licences may well have been 'legal' when they were enacted, which means that some of the licences issued under those *perda* probably remain legally valid today. To explain this, we need to discuss Government Regulation 6 of 1999 on Commercial Forestry Enterprises and Harvesting of Forest Products from Production Forests (the 1999 Government Regulation), issued soon after the fall of Soeharto as part of what appeared to be a genuine desire from the central government to satisfy regional demands for greater control over natural resources and the proceeds of their exploitation. The Regulation covered forestry exploitation concessions (*hak pengusahaan hutan*) and forestry product harvesting concessions (*hak pemungutan hasil*) (Article 4). Forestry exploitation concessions permitted rights holders to, in natural forests, fell trees, regenerate and manage forests and to manage and market forest products (Article 5(2)). Forestry exploitation concessions permitted planting, harvesting, management and marketing activities in plantation forests (Article 5(3)).

Critically, both governors and national ministries were granted authority to issue forestry exploitation concessions, although governors were limited to issuing rights over forest areas less than 10,000 ha (Article 11(2)).[5] For areas less than 50,000 ha, concessions could be granted through a simple application process (Article 7(2)). For larger areas, however, an 'auction' mechanism was required, under which the government would first designate the forest area for exploitation and then 'sell' that right to the highest bidder (Article 7(1)).[6] The national government was required to 'take into account' the views of the governor of the province when issuing these larger concessions (Article 11(1)).

---

5 Both national and regional governments were prohibited from issuing rights to a single entity over 100,000 ha per province, or over 400,000 ha throughout Indonesia (Article 8(1)(a)–(b)). In Irian Jaya, however, entities were allowed to hold up to 200,000 ha (Article 8(1)(c)).

6 For natural forests, concessions could last up to 20 years; for plantation forests up to 35 years (Articles 15(1) and 15(2)). Foreign legal entities were able to obtain rights over plantation land, but not natural forest (Article 10(2)(d)).

By contrast, the Regulation authorised country heads (or regents) to issue forest 'harvesting' concessions (Articles 1(17), 22(1) and 22(3)), apparently without the need for competitive processes.[7] The concession could be awarded over a 100 ha area for one year (Article 24(1)). It could also, however, simply cover 'a certain amount' (Article 24(1)(b)). Presumably this allowed exploitation until a particular volume of forest products was reached, regardless of how long this took, but the Regulation did not prescribe a limit to this amount. It is possible, then, that many of these licences remain valid, provided that the specified 'yield' has not yet been met.

Under the 1999 Regulation, district governments issued thousands of timber extraction permits and forest conversion licences. To that end, *perda* were often issued to provide a regional legal basis for these concessions and were also often the legal form used to grant the concession.[8] Even though these permits were intended to allow for small-scale logging operations, many local governments were said to 'abuse' these new powers, issuing thousands of concessions for large-scale logging over areas far exceeding these 10,000 and 100 ha limits. Some local officials were said to have granted multiple permits simultaneously, thereby opening large areas for logging, sometimes overlapping with concessions issued by the Forestry Ministry (Indrarto et al. 2012: 27–8).[9] Foreign entities, formally unable to obtain most types of these concession, reportedly obtained access by partnering with local concession holders.[10]

With logging spiralling out of control (Indrarto et al. 2012: 29), the central government sought to wind back the concession-granting powers of local governments. In 2000, the Forestry Ministry sought, by Decree,[11] to defer the implementation of a Decree it had issued in 1999[12] to 'implement' the 1999 Government Regulation. This was closely followed by another Forestry Ministry Decree in 2000,[13] which expressly prohibited

---

7 Although it appears that wholly or partly foreign-owned companies were not permitted to obtain extraction concessions (Article 22(2)).

8 See, for example, North Luwu District *Perda* 5 of 2001 on Licencing of Forestry and Plantation Enterprises in Luwu Utara, which created 19 types of forestry and plantation permit (Barr et al. 2006: 47). See also the various *perda* on the Forestry Ministry's website: www.dephut.go.id.

9 See Barr et al. (2006: 89) for statistics on the proliferation of small-scale licences during this time and the different labels attributed to the licences.

10 Also, as Indrarto et al. (2012: 27) observe, local governments were issuing concessions of a similar size to those of the national government, but without being subject to the Indonesian Selective Cutting and Planting (TPTI) system, which imposed various conditions, including requiring sustainable logging.

11 Forestry and Estate Crops Ministry Decree 310/KPTS-II/1999 on Guidelines for Issuing Forestry Product Harvesting Permits. This Decree confirmed that district-level governments have power to issue these permits and outlines the application process for obtaining them (Articles 4(1) and 4(2)).

12 Forestry Ministry Decree 084/Kpt-II/2000 on Deferment of the Implementation of Ministry of Forestry and the Estate Crop Ministry Decree 310/Hpts-II/1999 on Guidelines for the Granting of Extraction Permits.

13 Forestry Ministry Decree 05.1/Kpts-II/2000 on Criteria and Standards for Forest Product Use Permits and Forest Product Extraction/Harvesting Permits in Natural Production Forests.

governors, regents and mayors from issuing these permits. However, the Decree provided that existing concessions would remain in force (Article 4(a)).

The Forestry Ministry's regulatory efforts were largely ignored, with regional governments continuing to issue their own regulations and policies to maintain power to grant logging and forest conversion licences (Barr et al. 2006: 91). For example, the Berau district government, located in East Kalimantan, created its own type of licence: to Extract Timber from Private Land (*Izin Pemungutan Kayu Tanah Milik*, or IPKTM), which allowed logging and extraction outside the formally demarcated boundaries of forest areas. The Berau government also encouraged the registration of traditional land as private land so that these licences could be awarded over it (Barr et al. 2006: 91–2).

The central government responded by issuing Government Regulation 34 of 2002 on Forestry Administration and the Formulation of Plans for Forest Management, Forest Utilization and Use of Forest Estates (the 2002 Regulation), which replaced the 1999 Regulation (see Article 100).[14] This gives regents and mayors the power to issue harvesting permits for timber and non-timber products and exploitation concessions for non-timber products over forests located in their districts or cities; governors to issue such permits over forests that cross districts or cities; and the Minister to issue them over forests that cross provinces (Articles 38 and 40).[15]

However, the Regulation returns exclusive power to the Minister to issue timber exploitation concessions over natural and plantation forests in provinces, cities and districts, although this should be 'based on the recommendation' of the regent or mayor and governor (Article 42). The effect of the 2002 Regulation was, therefore, to 'effectively [recentralise] regulatory control over the harvesting, processing, and marketing of forest products, [in] particular timber' (Barr et al. 2006: 32). These concessions, awarded by way of auction (Article 43(3)–(4)), similar to the process described earlier in the 1999 Regulation, allow their holders to log, transport, process and trade timber products in natural forests (Article 29(1)) and to log and harvest plantation forests (Article 30(1)). The duration and geographic scope of the concession depends on the type of concession (see Article 35). For example, Article 35(3) states that a commercial use permit to exploit timber products in natural forests can be granted for a maximum of 55 years, whereas in a plantation forest a commercial use permit to exploit timber products can be granted for a maximum of 100 years (Article 35(5)). Many district governments appear to have followed the Government Regulation and no longer issue such licences (Barr et al. 2006: 104).

14 Similarly, Forestry Ministry Decree 6886 of 2002 on Guidelines and Method of Granting Forest Product Extraction Concessions (IPHH) which appears to implement Government Regulation 34 of 2002, abolishing the implementing regulation of the 1999 Regulation: Forestry and Estate Crops Ministers Decree 310/KPTS-II/1999.

15 The Regulation sets out different types of permits and concessions that can be issued to various entities, such as individuals, cooperatives or commercial enterprises. For a useful summary, see Barr et al. (2006: 50–1).

National-level conflict

Disputes between national institutions have also increased since Soeharto's fall. It is quite common for national ministries to clash over their relative jurisdictions and for national institutions, including ministries, to issue inconsistent laws or attempt to reach beyond their formal jurisdictions. The Forestry Ministry remains the most influential and powerful ministry in forestry matters and claims ultimate authority over forest management (Lang 2011). However, many other national government institutions and officials have, or claim to have, legitimate authority over forestry-related issues that bear on REDD+. The Moratorium, for example, directs the Forestry, Internal Affairs and Environment Ministers, the Heads of National Land Affairs, Spatial Planning, Surveys and Mapping Coordination Institutions, and members of the REDD Taskforce to help prevent deforestation. To this list, the Ministry of Energy and Mineral Resources, the Ministry of Agriculture and the powerful Ministry of Finance should be added. The result is a tangled web of overlapping and contested jurisdictions.

We provide two examples of the types of dispute that have arisen. The first is the apparent conflict between Forestry Ministry Decree 36 of 2009 on Permit Procedures for Carbon Sequestration and/or Storage Enterprises in Production and Protection Forests and Law 17 of 2003 on State Finance. Schedule III to the Decree specifies how proceeds derived from carbon sequestration and storage are to be shared between the State, the affected community and the project developer. The Finance Ministry has objected to this Decree, arguing that it, and not the Forestry Ministry, has the authority to manage the finances of the State and, therefore, to determine the 'split'. The legal ground on which the Finance Ministry claims authority is Article 6(2)(a) of Law 17 of 2003 on State Finance, which states that Presidential power to manage State finances is delegated to the Finance Minister.[16] Clearly, the Finance Ministry is of the view that the dividing of proceeds in Schedule III constitutes an exercise of this financial management power. In response to these objections, the Forestry Ministry announced plans to revise the Regulation in 2011, but to our knowledge, no changes had been made at time of writing.

Another example of an apparent conflict is between Agrarian Affairs Ministerial Regulation 37 of 2007 concerning Community Forests and the 1999 Forestry Law. The 1999 Forestry Law seeks to recognise the rights of traditional communities in several provisions.[17] It requires that the State 'pay regard to the rights of traditional communities', provided that doing so does not conflict with the national interest (Article 4(3)). Communities living

---

16 Although Ministers and heads of other institutions such as the Forestry Ministry are given authority over certain aspects of State finances, that authority is as a 'user' and not as a manager (Article 6(3)). The Law defines 'State finances' as 'all of the state's rights and obligations that can have a monetary value and all things in the form of money or goods that can become state assets related to the application of those rights and obligations'.

17 Traditional rights and REDD are considered in more detail in Chapter 9.

in and around forests also have rights to compensation for lack of access to forests resulting from demarcation of forest lands (Article 68(3)) and a right to compensation for loss of land ownership due to its designation as a forest area (Article 38(4)). Article 5 of the Law also recognises *adat* forests – regions within State forests where traditional communities can exercise particular rights, such as collecting forest products for their daily needs and managing forests in accordance with *adat*, provided their activity does not violate the law (Article 67).

However, the Agrarian Affairs Ministerial Regulation does not permit *adat* forest use in line with the Forestry Law. Rather, it allows Minister to grant two types of licence to forest communities. The first is to extract, for non-commercial purposes, up to 50 cubic metres per year from production forest zones that are not subject to other concessions (Articles 7 and 17). The second type is a licence to plant and harvest, provided there are no other concessions in force over the land (Article 22).

Of course, this leaves the entitlements of *adat* communities dependent on whether government officials follow the Forestry Law or the Agrarian Affairs Minister Regulation. The conflict also raises the question about what citizens can 'do' about inconsistent laws. Could a traditional community challenge the Agrarian Affairs Minister Regulation in an effort to have it invalidated? We consider judicial review of regulatory instruments later.

Rules of the game for resolving jurisdictional disputes – the hierarchy of laws

The starting point for resolving conflicting laws in Indonesia is the 'hierarchy of laws' (*Tata Urutan Peraturan Perundang-undangan*). This hierarchy, contained in Article 7(1) of Law 12 of 2011 on Lawmaking, is as follows:

a   The 1945 Constitution (*Undang-undang Dasar 1945*).
b   Decrees of the People's Consultative Assembly (*Ketetapan MPR*).
c   Statutes/Interim Emergency Laws (*Undang-Undang/Peraturan Pemerintah Pengganti Undang-Undang*).
d   Government Regulations (*Peraturan Pemerintah*).
e   Presidential Regulations (*Peraturan Presiden*).
f   Provincial Regulations (*Peraturan Daerah Propinsi*).
g   District/City Regulations (*Peraturan Daerah Kabupaten/Kota*).

In essence, each type of law must not conflict with any law higher than its own type in the hierarchy; and one type of law can amend or revoke a law lower than its own type in the hierarchy. So, for example, a provincial regulation will be legally valid – at least formally – only if, when passed, it does not contradict a Presidential regulation, government regulation, statute, People's Consultative Assembly Decree or the Constitution; and, once passed, it is susceptible to being overridden by any of those higher level instruments.

Many lower level laws are said to derive their legal legitimacy from a higher level law. For example, many government regulations are issued in response to a statutory provision that directs the government to issue that regulation, to provide further details about a matter covered very generally in the statute. It is common for years to pass before such 'implementing regulations' are issued, if they are issued at all. Current practice is that most legislative provisions that anticipate further regulations lie dormant until the regulations are passed – the courts will very rarely, if ever, 'fill in the gaps' left by statutes.

Article 8 of Law 12 of 2011 on Lawmaking reads:

(1) Types of laws other than those referred to in Article 7(1) include regulations stipulated by the People's Consultative Assembly, the People's Representative Council, the Regional Representative Council, the Supreme Court, the Constitutional Court, the National Audit Agency, the Judicial Commission, Bank Indonesia, Ministers, or equivalent agencies, institutions or commissions established by statute or by the government as required by statute, Provincial People's Representative Councils, Governors, District/City People's Representative Councils, the Mayor/District Heads, Village Heads or equivalent.

(2) The types of laws referred to in Article 8(1) are recognised and have binding legal force provided that they are required by higher level laws or on the basis of authority.

In practice, the hierarchy does not function adequately as a mechanism to resolve jurisdictional disputes between institutions and levels of government for several reasons. The first is that many types of law very frequently employed in Indonesia are not mentioned on the list contained in Article 7(1), so even though they are given legal force by virtue of Article 8(1), their authority vis-à-vis other laws is unclear. Classic examples are Ministerial regulations or decisions. They are among the most commonly used national regulatory instruments, allowing Ministers to regulate matters falling within their portfolios without needing to seek parliamentary or other government approval. The only laws dealing specifically with REDD+ are Forestry Ministry regulations.

What, then, is the status of Ministerial regulations? Because the hierarchy does not provide a clear answer to this question, different actors have voiced different views. On the one hand, some local government officials give them no weight at all, arguing that, because they are not mentioned in the hierarchy, they are not sources of law recognised within the Indonesian legal system (Indrarto et al. 2012: 28). This argument appears incorrect given that Article 8(1) states that types of law stipulated by Ministers 'have binding legal force provided they are required by higher level laws'. On the other hand, in response to these local government claims that Ministerial regulations have no legal force, the Minister of Justice and Human Rights reportedly issued a directive declaring that Ministerial decrees trumped regional regulations (Barr et al. 2006: 99). However, the correctness of this view is also doubtful because Ministerial regulations are not listed in the hierarchy. Further, it might be tempting to position Ministerial regulations between Presidential regulations and provincial regulations on the basis that Ministers are appointed by the President and they are senior national officials. However, this approach does not hold from a democratic perspective, because Ministers are appointed, not directly elected – whereas local legislatures and executives are directly elected by citizens. Also, if a Ministerial regulation and, say, a *perda* are both issued in response to a statutory delegation of power, should they not have the same authority? Uncertainty also arises about the types of law issued by the President, namely, Presidential Regulations, Decrees and Instructions. The hierarchy clarifies neither the relative status of these laws nor their status vis-à-vis other types of law.

Additionally, the hierarchy does not explain how conflicts between laws of the same type can be resolved. In Indonesia, principles of statutory interpretation common to most legal systems exist to help resolve inconsistencies between statutes. The first is *lex specialis derogat lex generalis*. According to this principle, if two inconsistent laws are applicable, the more specific over-rules the law of more general application (Hamzah 1986: 352). The second is *lex posteriori derogat lex priori*. Under this rule, if two laws conflict with one another, then the more recently enacted law prevails (Hamzah 1986: 352). However, these principles seem to operate primarily to help resolve disputes over laws that emanate from a single source: the national parliament. It is unclear whether these principles can be applied to resolve inconsistencies between laws issued by different institutions or levels of government. If, for example, two different Ministers issued contradictory laws, could the latter implicitly repeal the former? Or would the more specific prevail? To our knowledge, these issues have never been determined by an Indonesian court. Ultimately, however, the confusion arising from the hierarchy is largely moot. Unfortunately, most types of law on the hierarchy appear not to be 'enforceable' or 'reviewable' as against many of the others. Additionally, the Indonesian Supreme Court appears to shirk from even those conflicts that it has the power to review. We return to discuss the Supreme Court's judicial 'enforcement' of the hierarchy later. First, we consider the only other available review mechanism: central government review of *perda*.

Bureaucratic review

The 2004 Regional Government Law gives power to the central government to review *perda*. For most types of *perda*, local governments need not consult with the central government during drafting or before enactment; the central government review takes place after the local lawmaker has enacted the *perda*. Regional lawmakers must send their *perda* to the central government within seven days of enactment. The central government is to review the *perda* against two criteria: whether the *perda* breaches the 'public interest' (*kepentingan umum*) or contradicts a 'higher law' (*peraturan perundang-undangan yang lebih tinggi*) (Article 145(2)). This is a reference to the hierarchy of laws just mentioned.

If the central government considers that the *perda* breaches either of these criteria, the 2004 Autonomy Law allows the central government to invalidate it by Presidential Regulation (Article 145(3)). If, however, the government finds no fault with the *perda*, then it remains in force (Article 145(7)). Significantly, the central government's right of review expires after 60 days (Article 145(3)) so if it does not invalidate the *perda* within this time, then the *perda* continues in force by default (Article 145(7)). If the central government decides to invalidate, it must inform the relevant regional head – whether provincial, district or municipal – who is to prevent the *perda* from being further implemented or enforced and must revoke the *perda* within seven days (Article 145(4)). If dissatisfied with the central government's decision, the local lawmaker can lodge an application for review with the Supreme Court (Article 145(5)), discussed later. Different procedures apply for the review of *perda* that set local government budgets, impose regional taxes or user charges or relate to spatial planning. These types of *perda* need central government pre-approval.[18]

Between 2002 and 2009, the Home Affairs Ministry annulled 1878 *perda*; and in 2009–12 the Ministry cancelled 951 after reviewing 15,000 (Parlina and Aritonang 2013). By 2008, the Finance Ministry alone had, from 7200 *perda* received, recommended the revocation of around 2000, most of which sought to impose an illegal tax or user charge (Rosdianasari et al. 2009: ix). The Home Affairs Minister stated that his Ministry would review more than 2500 by-laws, both old and new, in 2013 (Parlina and Aritonang 2013). Despite the apparently large numbers of *perda* being reviewed or revoked, however, for several reasons the review process has not worked effectively – if it works at all – for almost all types of regional law. First, concerns have been raised that the central government allows through too many *perda*, that, on their face, contradict central government laws or breach fundamental human rights standards (Lewis 2003: 178), simply because the central government lacks the capacity to review all of the *perda* it receives. In this context, failure to revoke occurs because the mammoth task of reviewing all *perda* produced by Indonesia's 1000 or so regional lawmakers is conducted by relatively small teams that meet only weekly and do not allocate sufficient time to the task (Ray 2003: 18). The only category of regional laws that have attracted the attention of the central government are those imposing taxes or user charges in violation of national laws (Butt and Parsons 2012). Almost all *perda* invalidated thus far under the bureaucratic review process have fallen into this category.

18 For detailed discussion, see Butt and Parsons (2012). See Articles 185–9 of the 2004 Autonomy Law; Internal Affairs Ministry Regulation 53 of 2007 and Law 28 of 2009 on Regional Tax and User Charges.

Second, some regional governments are said to be failing to convey their new *perda* to the central government. This is hardly surprising given that no sanctions exist for non-compliance, except for the failure to submit *perda* imposing taxes and user charges for pre-enactment review. Estimates vary, but in the earlier days of decentralisation it seems that somewhere between only 30 and 40 per cent of *perda* were being sent to the central government (Lewis 2003).

There are, to our knowledge, no additional forms of 'bureaucratic review' under which other types of law can be reviewed as against higher level laws. Only the Indonesian courts can determine the legality of such laws by exercising judicial review, to which we now turn.

Judicial review by the Supreme Court

The Constitution grants the Supreme Court power to review legal instruments below the level of statute to ensure their compliance with statutes (Article 24A(1) of the Constitution). This jurisdiction is confirmed both in Law 4 of 2004 on the Judiciary (Article 11(2)(b)) and Law 14 of 1985 on the Supreme Court (Article 31(1)).[19] The Court has held this power since the 1970s (Lev 1978).

These Laws authorise the Supreme Court to annul a legislative instrument below the level of statute if: (a) it conflicts with a higher legislative instrument; or (b) the process by which it was enacted did not conform to legislative requirements (Article 31(2) of Law 14 of 1985). Legislative instruments struck down by the Supreme Court will no longer be legally binding (Article 31(4)) and no appeals against its review decisions are allowed.[20] These review applications can be made directly to the Supreme Court or via the applicant's local district court.[21] To have legal standing, an applicant must show that his or her rights or interests have been adversely affected by the enactment of the legal instrument (Article 31A(2) of Law 14 of 1985). Applicants may be individual Indonesian citizens, customary community groups, as well as public or private legal entities (Article 31A(2)). This avenue of review, therefore, is available to local governments and businesses alike. It is also the only means by which citizens can challenge lower level laws that do not comply with national legislation (Hukumonline 2006).

The 2004 Regional Government Law provides the Supreme Court with a narrower additional review power. Article 145(5) of the 2004 Law provides that a local lawmaker can, if dissatisfied with the central government decision to order revocation under the bureaucratic review process, lodge an application for judicial review with the Supreme Court (Article 145(5)). Presumably,

19 Law 14 of 1985 on the Supreme Court has been amended twice, first, by Law 5 of 2004 Amending Law 14 of 1985 on the Supreme Court, and, second, by Law 3 of 2009 Amending for the Second Time Law 14 of 1985 on the Supreme Court.
20 Article 9 of Supreme Court Regulation 1 of 2004 on Material Review.
21 Article 11(3) of Law 4 of 2004 on Judicial Power; Article 31(3) of Law 14 of 1985.

in these cases, the Court can consider whether the revoked local law was, in fact, contrary to a higher law or the public interest and reinstate the local law if it were not. This is the sole avenue for local legislatures and executives to challenge the central government's revocation of their laws. This type of review appears to be distinct from the Supreme Court's more general review power because it does not involve a lower level law being reviewed as against a 'statute'. Rather, the revoked *perda* is reviewed against the instrument that revoked it: either a Ministerial or Presidential decision.

There are very significant gaps in the judicial review jurisdiction held by Indonesia's courts. In summary, the Constitutional Court can only review statutes against the Constitution. The Supreme Court may only review lower level laws against statutes (under its general judicial review powers) and Ministerial or Presidential decisions that purport to invalidate *perda* (using its jurisdiction under the 2004 Regional Autonomy Law). There seems to be no judicial avenue to challenge lower level laws against other lower level laws mentioned on the hierarchy – such as Presidential regulations and government regulations – or even as against the Constitution. So, for example, there appears to be no mechanism for review of the 2000 Forestry Ministry Decrees,[22] mentioned earlier, which purported to postpone and then prohibit local governments from issuing various licences, as against Government Regulation 6 on 1999 on Commercial Forestry Enterprises and Harvesting of Forest Products from Production Forests, which appeared to *permit* the issuance of those licences. This is an important shortcoming in the context of REDD+ which was, at time of writing, regulated almost exclusively by lower level laws – among them government regulations, Presidential Regulations and Ministerial Decrees. There are simply no legal mechanisms available to review the compliance of local laws with the various relevant Forestry Ministry Decrees (see further Chapter 6).

Likewise, it seems that even though district/city *perda* fall below provincial *perda* on the hierarchy of laws, the Supreme Court can only review district/city *perda* against statutes. Indonesian law also provides no guidance about how to resolve conflicts between legislation from different provinces, such as might occur if two *perda* from neighbouring provinces purport to regulate trade between those two provinces or the management or use of forests that cross provincial borders. Further, the Supreme Court appears to lack jurisdiction to resolve conflicts between laws issued by various national government departments and ministries. Again, the Supreme Court can only review these lower laws as against statutes. The result is that, for the most part, conflicting laws simply remain on the books.

Even those conflicts that do fall within the Supreme Court's judicial review jurisdiction have rarely been resolved satisfactorily, with the Court appearing

22 Forestry Ministry Decree 084/Kpt-II/2000 on Deferment of the Implementation of Ministry of Forestry and the Estate Crop Decree 310/Hpts-II/1999 on Guidelines to the Granting of Extraction Permits.

reluctant to invalidate regulations, if it exercises its review power at all.[23] The Supreme Court has had, as mentioned, power to review lower level laws since the early 1970s, but it turned down all requests to exercise that power until after Soeharto's resignation in 1998 (Hoesein 2009). More recently, the Supreme Court has heard approximately 50 judicial review cases per year since 2004, under both its general review jurisdiction and its narrower jurisdiction granted under the 2004 Regional Government Law. In 2004–09, 200 judicial review cases were lodged with the Supreme Court (Parsons and Mietzner 2010: 190); in 2010, 61 cases (Mahkamah Agung 2010); in 2011, 50 cases and in 2012, 56 cases (Mahkamah Agung 2013). However, to our knowledge, the Court does not publish statistics about the outcomes of these cases. Judicial review cases make up a relatively small proportion of the Court's caseload, which was, at time of writing, around 13,000 cases per year (Mahkamah Agung 2013).

The Supreme Court has not, in the cases involving reviews which we were able to obtain and examine, actively exercised its judicial review powers as the Constitutional Court has. As some of the cases discussed in Chapter 9 show, the Constitutional Court often issues decisions that are discursive and reasoned, indicating that the Court has given consideration to the arguments put forward by the parties and providing justification for upholding some arguments but not others. By contrast, the Supreme Court is not renowned for this. In a significant number of cases in which the Court has reviewed *perda*, it has upheld the *perda* without indicating in its judgment that it has considered the content of the *perda* in any detail, or providing reasoned consideration as to whether the *perda* contradicts any higher laws. In the eyes of the Supreme Court, it seems that regional governments should and do have a broad discretion to pass laws to implement a very wide range of policies. To illustrate the Court's approach, we now turn to provide three case examples, all of which were Supreme Court reviews of *perda*.

The first is Supreme Court Decision 03 G/HUM/2002. The applicants, who were village heads, challenged Madiun District *perda* 4 of 2001 on

---

23 Until recently the Supreme Court had, by Supreme Court Regulation, imposed on applicants a 180-day time limit to lodge an application for judicial review of a law, running from the date that law was enacted (Supreme Court Regulation 1 of 2004, Art 2(4); Supreme Court Regulation 1 of 1999, Art 5(4)). The 180-day time frame was a highly restrictive constraint on judicial review and served primarily to lessen the Supreme Court's workload. The Supreme Court strictly observed this time restriction in most cases (see, for example, MA Decision 04 P/HUM/2000). The effect of this time limit was often to render regulations immune from review, particularly in the hands of savvy government officials. The government could, for example, simply ensure that the law remained undetectable for more than 180 days, such as by not making it publicly available or by refraining from implementing or applying it. The Court would often disregard arguments by applicants that they had been unable to file their application in time as they were not made aware of the existence of a new law (see, for example, Supreme Court Decision 05.G/HUM/2001). The 180-day time period was abolished by Supreme Court Regulation 1 of 2011. This is a significant advance, because it now opens the door for judicial review of older problematical laws, including *perda*.

the Establishment of the Village Representative Body (VRB). This *perda* purported to provide greater representation to women and youth in VRBs and, according to the applicants, in so doing breached MOHA Decision 64 of 1999. The Supreme Court did not delve into the merits of the arguments put forward by the applicants. It did not consider the substance of the *perda* or the MOHA Decision to respond to a preponderance of arguments put forward by the applicants. Rather, the Court upheld the *perda*, merely declaring that:

> [T]he contents of the *Perda* regulates the Region and is within its jurisdiction on the basis of Regional Autonomy … the contents of the *Perda* represent an implementation of, and are based on, Law No. 22 of 1999 on Regional Government.

The Supreme Court upheld the *perda* on an additional ground. By way of background to this argument, an Indonesian law, in its opening paragraph, usually lists other laws relevant to it, including the Constitution, statutes and any other pertinent lower level laws. If the law is itself a lower level law, it will usually list any higher level laws that it purports to implement. The *perda* under review in this case did not formally declare that it was implementing the MOHA Decision. For this reason, the Supreme Court declared that the *perda* could not be said to conflict with the Decision. This argument is feeble in the extreme: surely a lower level law can conflict with a higher level law even if it does not mention that higher level law in its opening paragraphs. If this were not the case, then the hierarchy would be a nullity: to avoid it, all that lawmakers would need to do would be to simply not mention any higher level laws in the opening paragraphs of their laws.

The second example is Supreme Court Decision 24 P/HUM/2002, in which the budget of Nusa Tenggara Timur Province, contained in a *perda*,[24] was challenged by several citizens. Again, the Supreme Court did not consider the merits of the case and upheld the *perda*, holding that:

> [T]he Regions have the authority to regulate their budgets in accordance with the conditions in their region … the provisions in the budget, both outgoing and ingoings, are still in the form of a plan, the realisation of which still depends on the actual incomings and needs. Changes can still take place … the implementation of the budget is public policy, which cannot be reviewed through legal aspects.

In light of this statement, it is difficult to imagine the Supreme Court ever invalidating a *perda* that contains a regional budget. Such *perda*, it seems, will always fall within the policy prerogative of local governments – something the Supreme Court will not touch.

---

24 East Nusa Tenggara Provincial Regulation 15 of 2001, 2 January 2002, on the East Nusa Tenggara Provincial Budget for 2002.

In a third case, the official report of which is not available on the Supreme Court's website, the Court was asked to review a *perda* issued by the municipal government of Tangerang. This *perda*, which attempted to prohibit prostitution, drew significant controversy in Indonesia and internationally when it was employed to detain, on suspicion of prostitution, a waitress waiting for a bus home one evening. The Supreme Court was asked to review the *perda* against myriad national legislation, but refused to do so, simply declaring, by press release rather than by case report, that the Tangerang municipal government was free to enact such policies by way of *perda*.

Conclusion: implications for REDD+

In the early years of decentralisation, a senior official from the national Finance Ministry stated:

> Since regional autonomy was introduced, a phenomenon that has emerged is the tendency for regions to wish to regulate everything on the view that all objects and subjects within their territory fall within their jurisdiction and must, therefore, be subject to the wishes of the region as regulated in regional regulations. What happens next is a type of euphoria, where the region appears to no longer observe the applicable rules, including by enacting regulations which regulate issues outside of their jurisdiction.[25]

The same can be said of various national institutions – particularly Ministries – which have issued a plethora of laws in recent years. As this chapter has shown, one result is a proliferation of new laws from regional and national institutions, many of which appear to be inconsistent or contradictory.

Of course, governments everywhere push the boundaries of legality and in a properly functioning separation of powers system it often falls to the judiciary to determine formally whether an act is illegal and, if so, to remedy it. Most executives and legislatures, in fact, seek to expand their power through regulation and administrative action and simply by doing so inevitably test the limits of their formal authority. In no area of law is this more apparent than in natural resources, including forestry. As mentioned, government institutions, both national and regional, have been jostling for authority over natural resources in an attempt to secure some of the income – both legal and illicit – that this authority brings.

Bureaucratic review fails almost completely to 'pick up' *perda* regulating any subject matter – regardless of their potential perversity or injuriousness for citizens or business, or their inconsistency with higher level laws – provided that those laws do not seek to impose a tax or user charge. To our knowledge, no concessions issued or forestry licences established by *perda*

25 Cited in Ismail (2003: 87–8).

have been invalidated. The result is that these concessions stand until they expire, and licence-creating *perda* remain on the books, even if they contradict higher level laws – such as Ministerial Decisions or Regulations – and are, therefore, formally invalid. This is likely to cause significant uncertainty for investors, including in REDD+ projects, who may be confused if presented with two inconsistent laws that purport to govern an aspect of the project.

Unfortunately, there is rarely an available legal or judicial avenue to satisfactorily resolve these types of inconsistency. We argue that the main problem lies with the Supreme Court and its limited jurisdiction. The Supreme Court is, of course, not to blame for the gaps in its own judicial review jurisdiction. However, even in disputes over which it does have jurisdiction, it has largely failed to satisfactorily referee them. The result is significant confusion and real uncertainty over what the 'law' is.

References

Barr, C., I. A. P. Resosudarmo, A. Dermawan, J. F. McCarthy and M. Moeliono (2006) *Decentralization of forest administration in Indonesia: Implications for forest sustainability, economic development and community livelihoods*, Bogor, Indonesia: Center for International Forestry Research (CIFOR).

Bell, G. F. (2001) 'The new Indonesian laws relating to regional autonomy: Good intentions, confusing laws', *Asian-Pacific Law and Policy Journal*, 2(1): 1.

Booth, A. (2011) 'Splitting, splitting and splitting again: A brief history of the development of regional government in Indonesia since independence', *Bijdragen tot de Taal-, Land- en Volkenkunde*, 167(1): 31.

Butt, S. (2010) 'Regional autonomy and the proliferation of perda in Indonesia: An assessment of bureaucratic and judicial review mechanisms', *Sydney Law Review*, 32(2): 177.

Butt, S. and N. Parsons (2012) 'Regional taxes and user charges in Indonesia', *Sydney Law Review*, 34(1): 1.

Fitriani, F., B. Hofman and K. Kaiser (2005) 'Unity in diversity? The creation of new local governments in a decentralising Indonesia', *Bulletin of Indonesian Economic Studies*, 41(1): 57.

Hamzah, A. (1986) *Kamus hukum*, Jakarta: Ghalia Indonesia.

Hoesein, Z. (2009) *Judicial review di Mahkamah Agung RI: Tiga dekade pengujian peraturan perundang-undangan*, Jakarta: RajaGrafindo Persada.

Hukumonline (2006) 'Berangkat dari pembatalan perda privatisasi rumah sakit: Problem hukum pengujian perda (1)', *Hukumonline*, 22 June.

Indrarto, G. B., P. Murharjanti, J. Khatarina, I. Pulungan, F. Ivalerina, Justitia et al. (2012) *The context of REDD+ in Indonesia: Drivers, agents and institutions*, Bogor, Indonesia: CIFOR and ICEL.

Ismail, T. (2003) 'Kebijakan pengawasan atas perda pajak daerah dan retribusi daerah', in *Partnership for Economic Growth, Decentralization, regulatory reform, and the business climate*, Conference Proceedings in Jakarta, 12 August (on file with authors).

Lang, C. (2011) 'Indonesia: The three draft decrees', *REDD Monitor*, 12 January.

Lev, D. S. (1978) 'Judicial authority and the struggle for an Indonesian *Rechsstaat*', *Law and Society Review*, 13: 37.

Lewis, B. (2003) 'Tax and charge creation by regional governments under fiscal decentralization: Estimates and explanations', *Bulletin of Indonesian Economic Studies*, 39(2): 177.

Mahkamah Agung (2010) *Laporan Tahun 2010*, Jakarta: Mahkamah Agung.

Mahkamah Agung (2013), *Supreme Court Annual Report*, PowerPoint Presentation, 13 March (on file with author).

Parlina, I. and M. Aritonang (2013) 'After straddling proposal, ministry to review bylaws', *Jakarta Post*, 18 January.

Parsons, N. and M. Mietzner (2010) 'Sharia by-laws in Indonesia: A legal and political analysis', *Australian Journal of Asian Law*, 11(2): 190.

Ray, D. (2003) 'Overview/summary paper' in *Partnership for Economic Growth, Decentralization, regulatory reform, and the business climate*, Conference Proceedings in Jakarta, 12 August (on file with authors).

Rosdianasari, E. S., N. Anggriani and B. Mulyani (2009) *Dinamika penyusunan, substansi dan implementasi perda pelayanan publik*, Pustaka Sutra, Bandung.

Smoke, P. (2005) 'The rules of the intergovernmental game in East Asia: Decentralisation frameworks and processes', in World Bank, *East Asia decentralizes: Making local government work*, Washington, DC: World Bank.

8   Judicial and administrative relief and remedies

This chapter assesses the prospects for litigation relating to REDD+ issues in Indonesian courts. Cases that may arise include disputes between the State and traditional communities with customary entitlements to forests. They also include criminal matters arising from breach of REDD+ or REDD-related laws. The courts in which these disputes and matters are likely to be decided are the general courts, the administrative courts and the Supreme Court (we consider the Constitutional Court and its likely role in relation to REDD+ in Chapter 9).

Key issues that arise in relation to litigation over REDD+ in Indonesia include whether traditional communities are likely to have standing to bring such claims, including as members of class actions, and the civil remedies available to them under Indonesian 'tort' law. There are also systemic problems within the Indonesian judiciary that might affect the outcome of such cases: namely, corruption and competence. These problems are unlikely to be resolved in the foreseeable future because Indonesian courts, particularly the Supreme Court, have staunchly and largely successfully resisted reforms aimed at detecting judicial misconduct and improving judicial accountability.

This chapter then briefly discusses the administrative 'relief' available to citizens through the Ombudsman, including in REDD-related cases, before providing a preliminary assessment of Indonesia's Freedom of Information Law, which came into operation in 2010, and its enforcement.

Overview of the Indonesian judicial system and standing

Four branches of the judicature exist under the Indonesian Supreme Court (*Mahkamah Agung*): the general courts (*pengadilan umum*), the military courts (*pengadilan militer*), the religious courts (*pengadilan agama*) and the administrative courts (*pengadilan tata usaha negara*). From most of these courts, there are two levels of appeal. The first is to a provincial high court (*pengadilan tinggi*). The second is an appeal, referred to as cassation (*kasasi*), to the Supreme Court in Jakarta.

General courts

The vast majority of civil litigation and criminal prosecutions, including for breach of criminal provisions of Law 41 of 1999 on Forestry (the Forestry Law), are heard at first instance in the general courts. These courts have jurisdiction over any matter not falling within the jurisdictions of other courts. The general courts also house what are referred to as 'special courts' (*pengadilan khusus*), many of which employ ad hoc judges. These special courts include: commercial courts, which primarily hear intellectual property and bankruptcy law cases;[1] human rights courts, which adjudicate cases of genocide and crimes against humanity;[2] the industrial relations court, which hears particular types of employment-related disputes;[3] the fishery crimes court;[4] the taxation court;[5] and the anti-corruption court.

Administrative courts

Administrative courts of both first instance and appeal have jurisdiction to hear disputes between Indonesian citizens and the government. A person or civil legal entity who feels that their interests have been damaged by an administrative decision can ask an administrative court to invalidate that decision and can seek compensation and/or rehabilitation.[6] The grounds for invalidation are that the administrative decision breached either the applicable law or principles of good governance.

The administrative courts will only hear cases about 'administrative decisions'. Such decisions are defined as 'written determinations issued by an administrative body or official containing an administrative act, based on applicable law, which are concrete, individual and final, and cause legal consequences for a person or a civil legal entity'.[7] An administrative official's refusal to issue a decision when the official is required to issue one is also an 'administrative decision' for the purposes of Law 5 of 1986 on the Administrative Courts (the Administrative Court Law, Article 3).

Some instruments and documents are explicitly excluded from the definition of 'administrative decision' and cannot, therefore, be challenged in these courts. These are administrative decisions that constitute a civil legal act; have not yet been formally approved; are regulations with general application; were issued under the Criminal Code, the Code of Criminal Procedure or another criminal law; were issued based on a judicial investigation; relate

1 Established under Law 4 of 1998 on Bankruptcy in the wake of Indonesia's economic collapse from 1997.
2 Established under Law 26 of 2000 on the Human Rights Court.
3 Established by Law 13 of 2003 on Employment and Law 2 of 2004 on Settlement of Industrial Relations Disputes.
4 Established by Law 31 of 2004 on Fishery.
5 Established by Law 14 of 2002 on the Taxation Court.
6 Article 53 of Law 9 of 2004 amending Law 5 of 1986 on the Administrative Courts.
7 Article 1(9) of Law 51 of 2009 amending Law 5 of 1986 on the Administrative Courts.

to military administration; or are about electoral results (Article 2 of the Administrative Court Law).

To establish a claim under the first ground – that the administrative decision breached the applicable law – the applicant or plaintiff must point to a law that prohibits or is otherwise contrary to the administrative decision in question. This law could be a statute or some other form of law accommodated within Indonesia's legal hierarchy. For example, if an applicant sought to challenge the legality of a concession, licence or permit issued by the government using this ground, it must prove that the permit or its issuance violated or contradicted a particular law or regulation (Santosa et al. 2012: 191). It is on this basis that mining and agricultural permits issued by national or local governments might be challenged if they breach forestry or spatial planning laws (Santosa et al. 2012: 191).

The second ground is a breach of the principles of good governance. The Elucidation to Article 53(b) of the Administrative Court Law defines the principles of good government as legal certainty, orderly State administration, openness, proportionality, professionalism and accountability – as referred to in Law 28 of 1999 on State Administration is Clean and Free from Corruption, Collusion, and Nepotism.[8] The Elucidation to Article 3 of Law 28 of 1999 provides very broad definitions of these terms:

- 'legal certainty' is the principle in a law State under which the law, appropriateness and justice are prioritised in every State administrative policy
- 'orderly State administration' means order, harmony and balance by State administrators
- 'openness' is the right of the community to obtain, without discrimination, correct and honest information about State administration
- 'proportionality' means balance between rights and responsibilities of State administrators
- 'professionalism' means prioritising expertise based on codes of ethics and applicable laws
- 'accountability' means that every activity or outcome from the activities of State administrators must be justifiable to the community of the people.

Even though these principles appear to be defined very broadly, they are, in fact, difficult to establish because Indonesian courts tend to construe them narrowly (Santosa et al. 2012: 191–2).

Any person or legal entity whose interests have been damaged by an administrative decision has standing to bring a claim before Indonesia's administrative courts. Importantly, in some cases representative organisations, such as non-government organisations (NGOs), might also be granted standing under Law 32 of 2009 on Environmental Management. This Law gives community

---

8 Law 9 of 2004 amending Law 5 of 1986 on the Administrative Courts.

members the right to bring a class action to pursue their own interests, and/ or the interests of the community, if they experience damage resulting from pollution or other 'environmental factors' (Article 91(1)). The 2009 Law also allows an environmental organisation – that is, a legal entity that has preservation of the environment in its charter as a purpose of its establishment and that has performed activities in accordance with that charter for at least two years (Article 92(3)) – to bring claims in the interest of preserving the environment (Article 92(1)). Article 92(2) allows these environmental organisations to ask the court to order the government to take particular steps to alleviate or mitigate environmental damage, but prohibits monetary claims (except to reimburse expenses).

The principal limitation on administrative proceedings appears to be the 90-day time limit established by the Administrative Court Law. Article 55 of this Law states that a case can only be lodged within 90 days of either the impugned administrative decision being published or the applicant receiving the decision, if he or she is named in the decision. For those to whom the administrative decision is not specifically directed, but who are adversely affected by it, the 90 days run from the time their interests were affected.[9] If this time period elapses, then the administrative court can refuse to hear the applicant's case.

Class actions

Class actions can be lodged with Indonesian courts, including in cases that we expect will arise from the implementation of REDD+. The process is governed by Supreme Court Regulation 1 of 2002 on Class Action Procedures. A class action can be brought when a group of people becomes so numerous that it is not effective or efficient for each member to sue individually or even together in one suit (Article 2(a)). The facts or events for which the group members seek a remedy must be similar and the legal basis for their claims must be substantially the same (Article 2(b)).

The statement of claim must identify a representative of the class bringing the claim, and must clearly define the 'class' bringing the action, although the name of each member need not be individually listed (Articles 3(1)(a) and (b)). The statement of claim must, of course, also specify the allegations made by the class against the defendant (Article 3(d)). The class can be divided into sub-groups if some members suffer different losses to others (Article 3(e)). The compensation claim must be clearly stated, along with the mechanism or procedure for compensation to be distributed to all members of the class, if their claim is successful (Article 3(f)).

The class representative does not need to be approved by all members to represent the group in the class action (Article 4). All members must,

---

9 See Part V of Supreme Court Practice Note 1 of 1991 on Guidelines on the Implementation of Provisions of Law 5 of 1986 on the Administrative Courts.

however, be informed of key stages in the process, including when a court decides on the validity of the class action, when the claim is to be brought before a court, when the case has been decided and when the compensation is to be distributed (Article 7). After being informed about the action by the representative, members of the class can 'opt out' by completing a form attached as an appendix to the Supreme Court Regulation (Article 8(1)). If the compensation claim is successful, then the presiding judge must determine the amount to be paid by the respondent; which classes and sub-classes are entitled to that compensation; the mechanisms for its distribution; and the role of the class representative, such as to notify members of the distribution (Article 9).

Articles 71(1) and 71(2) of the 1999 Forestry Law grant to a community the right to bring a class action for illegal damage caused to forests, which has adversely affected the life of that community. Presumably, such a community could not rely on Article 71 to seek redress for REDD-related action or loss because REDD+ activities will generally be directed towards protecting forests rather than damaging them. Nevertheless, the provision might apply if forest damage occurred in forest areas allocated for REDD+ in breach of REDD+ conditions or regulations or if the damage was inadvertent. In any event, Article 71 does not disturb the general rights of citizens to bring a class action for loss under the 2002 Supreme Court Regulation, such as if they suffer loss as a result of being excluded from forest resources or being denied the financial distributions to which they are entitled under REDD+ regulations.

*The Supreme Court (*Mahkamah Agung*)*

Appeals from the general, military, administrative and religious courts can be heard on cassation (*kasasi*) by the Supreme Court, Indonesia's highest court, at least in non-constitutional matters. Cassation hearings, a feature of many countries following the civil law tradition, are similar to appeals within the common law system, but they are only concerned with legal aspects of the case; and their main purpose, at least formally, is to determine whether the lower courts have applied the law correctly to the case at hand. They are usually conducted 'on the papers' – that is, the parties submit their arguments in documentary form. There is usually no opportunity to submit oral argument or to directly challenge arguments made by the other party, except by way of these documents. However, even though the Supreme Court is theoretically not required to adjudicate on the facts of the case or on evidence produced by the parties, it commonly does so and can even call witnesses (Pompe 2005).

The Supreme Court has other important functions.[10] One is reviewing laws of a level 'lower' than national legislation – such as government regulations,

---

10 These are outlined in Article 24A of the Constitution.

Presidential Decisions and Ministerial Decrees – to ensure that their content is consistent with the legislation. The Court can also review the formal validity of lower level laws: that is, it can review the 'procedures used to enact the law'. However, the Supreme Court cannot review the constitutionality of statutes – this is the exclusive task of the Constitutional Court. The Supreme Court can also provide advice on 'pardons and rehabilitation' and supervises the lower courts, legal advisors and notaries. The Court also settles jurisdictional disputes between courts.

Appeals and enforcement

Civil cases[11]

Under Indonesian civil procedural law, an applicant who wishes to lodge an appeal against a district court (*pengadilan negeri*) decision must do so at the district court where the case was first heard, within 14 days of that decision being handed down or the parties being informed of it.[12] After this period, the parties are generally deemed to have accepted the decision. Consequently, appeal requests will usually be rejected and the original district court decision will become binding.

Dissatisfied parties can request a cassation (*kasasi*) hearing from the Supreme Court, mentioned earlier, provided that their case meets some minimal requirements. Articles 45A(1) and (2) of Law 5 of 2004 Amending Law 14 of 1985 on the Supreme Court (the 'Supreme Court Law') require the Supreme Court to accept cassation applications, unless the case is an appeal of:

- a pre-trial hearing
- a criminal case for a crime with a maximum imprisonment term of one year or less
- an administrative law case concerning a decision of a regional official, the reach of which is regional.

These grounds are extremely broad, allowing appeals of virtually all civil cases. The cassation request must be lodged with the district court that first heard the case, within 14 days of the appeal court decision being handed down (Article 46 of the Supreme Court Law). If the court does not receive the application within this period, the cassation request should be rejected.

Enforcement of Indonesian judicial decisions is notoriously complex and difficult. Under Indonesian law, a decision must generally be binding (*mempunyai kekuatan hukum tetap* or, literally, to 'have permanent legal authority') to be enforceable. A judicial decision is considered binding in two circumstances. The first is where it has been appealed to the Supreme Court. The

---

11 This section draws on Butt (2008).
12 Article 7(1) of Law 20 of 1947 on Appeals Courts.

second is where the time limits have lapsed for an appeal to a high court or for cassation by the Supreme Court. Lodgement of a request for an appeal or cassation will generally constitute a stay on enforcement.

If the unsuccessful party does not voluntarily comply with a 'binding' decision, the successful party must apply for a further order compelling the unsuccessful party to comply with the most recent judicial decision dealing with the dispute (Article 195(1) of the Code of Civil Procedure in force in Java and Madura (the HIR)).[13] The recalcitrant party is called before the chairperson of the district court that initially heard the case, which directs that party to comply with the decision within eight days or fewer (Article 196 of the HIR). If the unsuccessful party does not comply or does not attend the enforcement hearing despite being validly summonsed, then the court can take action to ensure compliance. For example, it can seize and auction property to finance damages awarded to the successful party; or it can put a monetary value on an act or service that the unsuccessful party refused to carry out (Articles 197(1), 200(1) and 225(1) of the HIR).

These provisions give rise to a number of problems, many of which cause or contribute to long delays in the enforcement of, and perhaps even eventual non-compliance with, court decisions at every level of Indonesia's judicial hierarchy. For example, the HIR does not require district court chairpersons to immediately respond to a request for an enforcement order – or even within a particular time. As a result, courts can allow such requests to languish deliberately or through administrative incompetence. Furthermore, the HIR does not impose sanctions on parties who fail to comply within the eight-day deadline for compliance.

As for administrative court decisions, litigants have frequently encountered problems when trying to have these enforced against the government. It has been common for government institutions to ignore the decisions of the administrative courts or to delay enforcement for as long as possible (see Bedner 2001). However, recent reforms appear to have given administrative court decisions more 'teeth'. Amendments to the Administrative Court Law enacted in 2004 impose fines and administrative sanctions on administrative decision makers who do not comply with administrative court decisions.[14] Amendments in 2009 provide that if an administrative court decision is not complied with within 60 working days, then the administrative decision complained of becomes no longer binding.[15] If the decision has not been implemented after 90 working days, then the plaintiff can seek an order from the administrative court commanding the public official to implement that decision. If the decision is not complied with, then the relevant official faces a fine.

---

13 Indonesia has two Codes of Civil Procedure – the *Herziene Indonesisch Reglement* (the HIR, which applies in Java and Madura) and the *Reglement Buitengewesten* (the RBg, which is in force in the rest of Indonesia).

14 Law 9 of 2004 amending Law 5 of 1986 on the Administrative Courts.

15 Law 51 of 2009 amending Law 5 of 1986 on the Administrative Courts.

Clearly, the enforcement powers of the administrative courts, although significantly improved by the 2004 and 2009 amendments, remain weak. It seems that an administrative decision, even if found to be illegal by an administrative court, can continue to be applied – at least for 60 working days – without adverse consequences for the official or government institution involved.

Criminal cases

Under the Code of Criminal Procedure (*Kitab Undang-undang Hukum Acara Pidana*, or KUHAP), an appeal to a high court against a district court decision must be lodged within seven days of the district court decision being handed down or being made known to a defendant who was not present when the decision was handed down (Article 233(2) of the KUHAP). If the defendant or prosecution does not appeal the decision within seven days, they are considered to have accepted the decision (Article 234(1)) and that decision becomes enforceable. It appears that the grounds for appeal are a very broad 'failure to correctly apply procedural law, mistake or incompleteness' (Article 240(1)). The high court can uphold the district court's decision or overturn the first instance court's decision and decide the case itself (Article 241(1)).

If dissatisfied with the high court's decision, the defendant or prosecution can lodge an appeal with the Supreme Court. A cassation application must be sent, by the applicant to the district court which initially heard the case, within 14 days of the applicant being informed of the appeal decision for which cassation is sought (Article 245(1) of the KUHAP). Again, if the cassation application is not lodged within 14 days, the parties are considered to have accepted the decision (Article 246(1)) and the decision becomes binding and enforceable.

Prosecutors are responsible for enforcing judicial decisions in criminal cases.[16] Enforcement is to be supervised by a judge (Article 277 of the KUHAP) who is to ensure that the decision is complied with (Article 280(1)).

Peninjauan kembali *(PK)*

The Supreme Court is authorised to reopen and review 'permanently binding' decisions from all of the courts from which it can hear appeals. It can even reopen and review its own decisions. This form of Supreme Court review is referred to as *peninjauan kembali* (PK). If reviewing one of its own decisions, the panel of judges presiding over the PK will be different to the panel that heard the original cassation application. Before 1970, the PK process was rarely used, being primarily employed to enable the reconsideration of

---

16 Article 270 of the KUHAP; Article 27 of Law 5 of 1991 on the Public Prosecution.

judicial decisions made during Dutch colonisation or under the Soekarno regime (Tasrif 1971). It is now so commonly employed by litigants that many lawyers now treat it as a *de facto* final judicial appeal.

In civil cases, the Supreme Court is formally permitted to conduct a review only in specified circumstances. These are where:

• the previous decision was based on subsequently discovered deceit of a party to the litigation or based on fraudulent evidence
• new determinative evidence was discovered after the case was decided
• a judge awards more than was sought by a party
• judges in similar cases involving the same parties, in a court of the same or a higher level, reached contrary decisions
• a clear judicial error has been made.[17]

A time limit of 180 days for lodgement, which runs from the time of discovery or occurrence of these events, is generally applicable (Article 69 of the 1985 Supreme Court Law).

In criminal cases, a PK can be lodged at any time after the decision (for which review is sought) has become binding and enforceable (Article 264(3) of the KUHAP)). The KUHAP provides three grounds for a PK:

1   If a new circumstance comes to light that gives rise to a strong suspicion that if that circumstance were known when the proceedings took place, the result would have been an acquittal, a dismissal of all charges, a rejection of the prosecutor's charges, or the application of a less serious criminal provision to the case (Article 263(2)(a)).
2   If, in one case, there is a declaration that something has been proved and which is used as a basis of the decision, but which conflicts with the evidence in another case (Article 263(2)(b)).
3   If the decision clearly indicates judicial error or a mistake (Article 263(2)(c)).

Unlike appeals and cassations, PK applications in criminal cases seem not to be subject to time limits.[18]

Remedies

Civil claims for 'unlawful acts': Article 1365 of the Civil Code

Article 1365 of the Civil Code states that a person who causes loss to another person by means of an unlawful act must, because of his or her fault

---

17 Article 67 of Law 14 of 1985 on the Supreme Court.
18 Article 264(3) of the KUHAP; see also Article 76 of the Supreme Court Law, as amended by Law 5 of 2004.

in causing the loss, compensate for that loss. The plaintiff must prove the elements of the claim, including the defendant's fault or culpability and a causal connection between the allegedly wrongful act and the damage suffered.

Article 1366 is related to Article 1365. It reads: 'Every person is responsible, not only for loss caused by their acts, but also for loss caused by negligence or lack of care.' Together, these two provisions provide the primary legal basis for Indonesian tort law, with Article 1365 applying to acts causing loss and Article 1366 applying to negligent omissions. Compensation can be in the form of money or restoration to the plaintiff's initial position, a declaration that the act was unlawful, prohibition from performing a particular act and/or an order requiring the defendant to perform a particular act (Agustina et al. 2012: 15; Ismail 2005: 45).

Article 1365 claims 'dominate' the workload of the general courts in civil matters (Agustina et al. 2012: 3; Fuady 2002: 1). The Civil Code was brought into force during the Dutch colonisation of Indonesia and remains mostly intact. Well over 150 years old, it is largely out of step with modern legal needs. Article 1365 is often used as a catch-all provision to remedy a variety of gaps in the Code. It has, for example, been used to enable plaintiffs to obtain damages against those who have acted in bad faith in patent[19] and trademark[20] violations, land disputes[21] and even in environmental matters. It also enables plaintiffs or victims to claim compensation for criminal acts (Fuady 2002: 21). Article 1365 has been used to provide damages in defamation and consumer protection cases, corporate crimes and banking offences (see Ismail 2005: Chapter III).

Article 1365's flexibility comes in part from the fact that the Civil Code itself does not define 'unlawful' act, which has left its meaning and interpretation to doctrine and judicial practice. Most Indonesian courts now interpret 'unlawful' more broadly to include merely inappropriate or immoral acts, judged by community standards, even if they are not explicitly prohibited by a written law. Clearly, using this interpretation means that Article 1365 can be relied on to establish a very broad ambit of potential civil liability. According to a publication issued by the government's National Legal Development Institute, an unlawful act is committed if it:

- contravenes another person's rights
- contravenes the legal obligations of the perpetrator
- conflicts with morality or appropriate behaviour
- conflicts with what is required for societal mixing with respect to people and property (BPHN 1993/94: 17).

19 Supreme Court Decision 99K/Sip/1976.
20 Supreme Court Decision 1954/Pdt/1987.
21 Supreme Court Decision 2939K/Pdt/1987.

We also note that Indonesian law recognises the 'unlawful act' concept within the criminal law. Use of the concept in criminal cases has, however, been viewed with disfavour in a recent Constitutional Court case about the legality of its use to convict defendants in corruption cases.[22] Its use is unlikely to arise in REDD-related cases because most acts that cause loss – for example, illegal logging, illegal issuance or breach of concessions – are likely to constitute 'crimes' under written laws, such as the 1999 Forestry Law, the 1999 Anti-corruption Law or even the general Criminal Code (*Kitab undang-undang Hukum Pidana*, or KUHAP). There will, therefore, rarely be a need to use a 'broad' interpretation of the unlawful act concept. However, if the government or an investor does perform an act that – while not illegal in the sense that it breaches a written law – incorporates an element of deceit, misrepresentation, concealment of facts, subterfuge, illegal circumvention, breach of trust or fraud, then it might arguably constitute an 'unlawful act' (Ismail 2005: 5) and that act might be criminally punishable.

Obstacles to judicial relief

Significant judicial reforms have been achieved since Soeharto left Presidential office in 1998. The courts are now institutionally separate from the government and regularly issue decisions in which the government is defeated. Yet this independence has itself given rise to new problems. Responsibility for judicial administration was transferred from the government to the Supreme Court as part of the *satu atap* ('one roof') reforms. This responsibility is indeed burdensome, with the Supreme Court now administering 67 appeal courts and 706 first instance courts (including maintaining buildings and infrastructure) and administering 34,652 judicial officers, of which 7780 are judges (Mahkamah Agung 2013). These tasks, it is often suspected, have distracted the attention of many Supreme Court judges from their key adjudicative duties, resulting in a reduction in the quality of decisions.

This institutional independence has also failed to produce improvements in judicial standards of competence or integrity.[23] There remains a widespread perception that the majority of Indonesian judges are corrupt, incompetent or both. Further, the judiciary, led by the Supreme Court, has used its new-found independence to put up strong resistance to other reforms designed to provide accountability mechanisms to counterbalance the courts' increased autonomy. We now turn to discuss judicial corruption and accountability in more detail.

---

22 For more on the use of the unlawful act concept in criminal cases and this Constitutional Court decision, see Butt (2009, 2012a).

23 For more on judicial competence, see, for example, the Supreme Court's own publication, which states: 'In addition to partiality, the problems plaguing the judicial sphere are complemented by low judicial competence ... [T]ruth and justice, which should normally [be] applied to legal issues, is seldom reflected in court rulings' (Mahkamah Agung 2012: 1).

Judicial corruption[24]

For decades, the Indonesian judiciary has been widely regarded by many Indonesians as one of the nation's most corrupt institutions. Surveys indicate that the reputation of the judiciary, ironically, is for its propensity to act illegally rather than its capacity to enforce the law, let alone deliver 'justice' (Asia Foundation and ACNielsen Indonesia 2001). Popular belief has it that most of Indonesia's judges and court officials are willing to accept, or even to extort, bribes from litigants to secure victory in their cases, with the Supreme Court seen as one of the most corrupt courts in the country. The Indonesian joke – which even judges will tell – is that the word *hakim* (judge) is short for *hubungi aku kalau ingin menang* (contact me if you want to win).

It is also often said in Indonesia that corruption within the judiciary is systematic and institutionalised: illicit payments are filtered into patronage networks, within which the recipient's superiors will take a percentage of the bribe. For these reasons, the Indonesian justice system is often described as a 'mafia' (*mafia peradilan*) because most bribes are made as part of a complex web of well-organised 'arrangements' involving a number of corrupt players, rather than just a few rogue individuals. Even Supreme Court parking attendants are said to be involved (ICW 2001).

Senior Indonesian judges, including retired Supreme Court Chief Justices, have admitted that there is much accuracy in these popular perceptions. Former Chief Justice Soerjono, for example, estimated that 50 per cent of Indonesia's judges were corrupt (Pompe 2005: 414), as did former Chief Justice Asikin Kusumaatmadja (Inside Indonesia 1997), although well-regarded lawyers have claimed that the proportion of corrupt judges is closer to 90 per cent (Inside Indonesia 1997). Another former Supreme Court Justice, Adi Andojo, went even further in the mid-1990s by publicly naming other Supreme Court judges whom he accused of receiving massive bribes from litigants (Tempo Interaktif 1997). These judges were, however, never pursued in relation to these allegations. The situation, according to one lawyer, is so dire that Indonesian courts:

> [S]houldn't be called a house of justice [*kantor pengadilan*] but instead an auction house [*kantor lelang*]. An auction house for cases ... the judge himself calls me or the prosecution calls me or a policeman calls me – they're the ones who ask me, 'Do you want your client to be helped or not? Does your client have enough for a donation or not?' So they're actively pursuing this. They're the producers and they're offering their wares ... 'Do you have money or not? If you don't, I'll make an offer to your adversary.' Now when I get a telephone call like this, I'm no longer surprised ... Now I feel that it is natural. (Pemberton 1999: 200)

24 This section draws on Butt and Lindsey (2010a).

In 2001, the respected NGO Indonesia Corruption Watch (ICW) conducted dozens of interviews with judges, lawyers, court employees, prosecutors, litigants and police. Many interviewees testified that bribery was very widespread and that they had either witnessed or participated in it. One interviewee even claimed: '[W]hile I have been an employee of the court there have been no cases which did not involve the payment of money.' He concluded that: '[I]f most people who take bribes go to hell, I think that no judges will be let in heaven.' Quoting from the ICW research, the World Bank (2003: 81) has noted in a report on corruption in Indonesia that:

> Judicial corruption is not ... considered anything unusual in Indonesia. Many lawyers feel no shame at offering sums of money to judges and attorneys in specific cases ... at the same time, there is no sense of shame for judges, attorneys, police and registrars to solicit money from lawyers. Whereas in the past corruption was handled secretly, it is now carried out openly. We need no longer talk about the courts as a place to find justice, but as a place to buy justice. Whoever pays the most will get the 'justice' that he/she wants.

The problem is so acute that President Susilo Bambang Yudhoyono established a Legal Mafia Eradication Taskforce (*Satgas Pemberantasan Mafia Hukum*).[25] This Taskforce operated for two years, until 30 December 2011. The main function of the Taskforce was 'coordination, evaluation, correction, and monitoring so that the eradication of the judicial mafia, especially within the police force, the prosecutor's office, the courts, and correctional institutions (law enforcement institutions), can be done effectively' (Legal Mafia Eradication Taskforce 2010: 4). However, as the Taskforce itself notes, this was no easy task because '[t]he practices of the judicial mafia in Indonesia have been going on for a long time and are carried out using many modi operandi that are increasingly sophisticated' (Legal Mafia Eradication Taskforce 2010: iii). To help it develop strategies to prevent and eradicate corruption in law enforcement institutions, the Taskforce researched the modus operandi of the legal mafia in various law enforcement institutions. According to a report produced by the Taskforce:

> The process that takes place just before the judges' deliberation and courtroom speech (reading out the decision) is the most vulnerable point. At this stage, the litigant or his/her advocates or through mediators/brokers often tries to offer a reward to the judge to decide the case in accordance with their expectations. Or a rogue judge, either directly or through a mediator, makes an offer or offers to one party or the advocate to rule favourably in their case by asking for some compensation, of course. More subtle ways that are sometimes carried out is that the rogue judge

25 By Presidential Decree 37 of 2009.

slows down the decision process as a sign that the party/defendant should contact the judge. In some cases, rogue judges even 'auction off' the decision, and he will declare the party who makes the highest payment the winner. In minor cases or cases where the legal issues handled are very clear, generally the judge will decide the case according to existing law but will ask the winning party for 'gratitude money' (or the winning party voluntarily gives a gift or money to express his/her gratitude).

It is also often the case that there are rogue prosecutors/advocates/ brokers in the court who deceive one party/parties (in civil cases) or the defendant/victim (in criminal cases) pretending that the judge is asking for a certain amount of money, but in fact the judge did not do so, or even if he asks for money the amount is not as high as stated. (Legal Mafia Eradication Taskforce 2010: 18–19)

In recent years, these perceptions of judicial corruption have been confirmed by successful prosecutions. In February 2012 the Jakarta Corruption Court sentenced Commercial Court Judge Syarifuddin Umar to four years in prison and ordered him to pay Rp 150 million in fines for accepting Rp 250 million in bribes while handling a bankruptcy case. In June 2011 former District Court Judge Muhtadi Asnun was sentenced to two years in prison for receiving US$40,000 in bribes from a defendant. In August 2010 Jakarta State Administrative High Court Judge Ibrahim was sentenced to six years in prison and fined Rp 200 million for receiving Rp 300 million in bribes.

Accountability mechanisms[26]

Unfortunately, it appears that the courts' competence and corruption problems are likely to remain for some time. Attempts at improving judicial transparency and accountability, particularly through external mechanisms, have been consistently thwarted by the Supreme Court itself.

Before the *satu atap* reforms of 2004, external monitoring of the courts was largely the responsibility of the government – primarily the Ministry of Justice – with the assistance of the Supreme Court. With the *satu atap* reforms came establishment of the Judicial Commission, which began operating in 2005. It is an independent body, with its seven members appointed by the President on the recommendation of the national parliament (Article 24B(4) of the Constitution).

Law 22 of 2004 on the Judicial Commission (the 2004 Law) requires the Judicial Commission to supervise the performance and behaviour of judges from all Indonesian courts, including the Supreme and Constitutional Courts, as part of its function to 'uphold the honour and dignity, and to ensure the [good] behaviour, of judges' (Articles 13(b) and 20).[27] In this context, Article

---

26 This description draws on Butt and Lindsey (2012).
27 See also Article 24B of the Constitution.

22(1) of the 2004 Law requires the Judicial Commission to investigate suspected breaches of proper judicial behaviour; call and seek explanations from judges suspected of breaching the judicial code of ethics; report the findings of investigations; and make recommendations to the Supreme Court and/or the Constitutional Court (and to send a copy to the President and the national parliament).

However, the supervisory powers of the Commission are limited, presumably in the interests of judicial independence.[28] If the Commission suspects a judge of acting improperly, it can propose that the judge be punished by written reprimand, suspension or dismissal (Article 23 of the 2004 Law). However, the Commission can take no further direct action – it can only send the proposed sanction and reasons for suggesting it to the Supreme or Constitutional Court leadership. Whether these allegations are pursued or sanctions are imposed is, therefore, almost entirely a matter for those courts.

The Judicial Commission has indicated a strong desire to perform its functions with some vigour. In its first year, it received 820 complaints and reports about judicial misconduct, called 74 judges to account for their actions and recommended that the Supreme Court take action against 18 judges. However, the Supreme Court has rarely responded to its requests to take action against particular judges, if at all. In fact, the Supreme Court has even resisted attempts by the Commission to investigate its judges.

In 2006, for example, members of the Supreme Court challenged, before the Constitutional Court, provisions of the 2004 Law that purported to allow the Commission to supervise judges, including by analysing its decisions. The Constitutional Court held that, for reasons of judicial independence, the Judicial Commission could not analyse judges' decisions to assess whether judicial misconduct has occurred (Butt 2007). The decision has left the Judicial Commission able to do very little beyond suggesting new appointments to the Supreme Court.

Apparently unsatisfied with neutering the Commission, the Supreme Court has also recently invalidated important parts of its own Code of Ethics that had drawn heavily on the Bangalore Principles of Judicial Conduct (2002). The provisions the Court invalidated had prohibited judges from making mistakes in their decisions, disregarding facts that could disadvantage a party, otherwise favouring a party or handling cases in which they have an interest. They also required judges to maintain and enhance their knowledge, skills and personal qualities; to respect the rights of parties; to understand and perform their tasks in accordance with the law so that they can apply the law correctly; and to meet their administrative responsibilities (Butt 2012b). The rationale for the Court's decision was, again, that these provisions had purported to allow the Judicial Commission to move beyond the supervision of mere behaviour to second guess the 'cognitive processes' that led to judicial

---

28 The following section draws on Butt and Lindsey (2010b) and Butt (2007).

decisions. With respect, it is difficult to see how requiring judges to avoid making mistakes and bias and to require them to maintain their skills and perform their tasks in accordance with the law, can possibly affect judges' decision making in particular cases.

Detecting and punishing judicial impropriety is now, in a formal sense, almost exclusively an internal matter for the Supreme Court, based on an Ethics Code that the Supreme Court itself has partly demolished. Yet the Supreme Court's performance in carrying out its supervisory role is highly questionable. Some Indonesian lawyers have claimed that, given its adjudicatory and administrative functions, the Supreme Court neither has sufficient time to carry out its supervisory role properly nor is it particularly interested in this role. According to a former Supreme Court judge, the Supreme Court's supervision of lower courts is almost completely non-functional, particularly with respect to supervision of the quality of judicial decisions. Even when some form of supervision does take place and a problem is identified, further action is rarely taken.

The Ombudsman

The National Ombudsman Commission (*Komisi Ombudsman Nasional*) was established by Presidental Decree 44 of 2000. In 2008 the Ombudsman was given a statutory basis (Law 37 of 2008). The Ombudsmen has seven members, including one chairperson and one deputy chairperson.

Its primary function is to monitor and receive complaints about the administration of public services at the national and local levels (Article 6). It can conduct its own investigations and seeks to cooperate with other law enforcement institutions. The Ombudsman could, therefore, receive complaints about REDD+ issues that arise out of perceived government maladministration and could investigate those complaints. However, its powers are limited to making reports, recommendations and complaints (Articles 35–8 of Law 37 of 2008).

The Ombudsman has received, and continues to receive, many complaints from citizens (Sherlock 2002; Crouch 2007).[29] However, it has been able to do very little to improve government or judicial impropriety. According to Butt and Lindsey (2012: 100):

> Though initially flooded with complaints, the Ombudsman is now considered 'toothless' as a check on government action and has become largely irrelevant to public life in Indonesia. A variety of reasons – such as insufficient budget and political support – are commonly cited as

---

29 In its first year alone, the Ombudsman received 1723 complaints, mostly about court and police corruption as well as maladministration (Crouch 2008: 386). After a sharp drop in complaints, it received 1237 official complaints in 2009 (up 17.23 per cent from 2008) and 1154 in 2010 (Ombudsman Republik Indonesia 2010: 20).

causes of its impotence (World Bank 2004: 13). The Ombudsman's main weakness, however, is its inability to compel government officials and departments to respond to its inquiries and recommendations. The 2008 legislation did not give it coercive powers and many officials simply ignore the Ombudsman's requests and findings (Sherlock 2002: 369–70). For the most part, the Ombudsman can only attempt to 'shame' impugned officials and government departments through the media and hope that they remedy the problem. This strategy has not proved particularly successful.

In practice, then, citizens and communities adversely affected by government maladministration in respect of REDD+ could complain to the Ombudsman, but this would be unlikely to lead directly to redress. Nevertheless, the threat of negative media coverage might, in some circumstances at least, prompt a government response.

Freedom of information

In 2008 Indonesia introduced its first 'freedom of information' statute – Law 14 of 2008 on Disclosure of Public Information (the FOI law)[30] – which became fully operational in 2010. This Law is an important component of the government transparency and accountability mechanisms established after Soeharto and his authoritarian 'New Order' government fell in 1998. This section briefly assesses the extent to which Law 14 has been effective in requiring public bodies to disclose 'public' information that they would rather keep within their ranks. We begin with a description of the key features of the FOI Law before discussing Information Commission and judicial decision making in freedom of information cases. We demonstrate that the Commission and the courts have ordered disclosure of requested information in the vast majority of reported cases.

The 2008 FOI Law

The FOI Law requires that 'all public information' be 'open and accessible' to 'users of public information' (Article 2(1)), which includes Indonesian citizens and legal entities. All citizens and entities have the right to request, view, understand, obtain a copy of and distribute public information.[31] All 'public bodies' (*badan publik*) must comply with this statute – that is, they must provide accurate, correct and clear information to members of the public who have requested information, unless that information falls within one of the 'excluded information' categories under the Act (discussed later). The

---

30 *Undang-Undang Nomor 14 Tahun 2008 Tentang Keterbukaan Informasi Publik.*
31 The FOI Law also gives the right for 'all persons' to 'attend public meetings to obtain public information': Articles 4(2) and (3).

Law specifies types of information that public bodies cannot exclude and therefore must disclose. These include judicial decisions; regulations; administrative decisions and policies; and orders to cease investigations or prosecutions (Article 18(1)). The public body must provide the information quickly, cheaply, in a simple manner (Article 2(3)) and in comprehensible language (Article 10(2)).

'Information' is broadly defined in Article 1(1) to include: any information, statement, idea or sign that has value, meaning, or a message – including data, facts or explanations – that can be seen, heard or read, whether in electronic or non-electronic form. 'Public information' means information produced, stored, managed, sent or received by a public body that concerns the public interest and relates either to the administration of the State or the administration of another public body (Article 1(2)).

Public bodies are defined broadly to include entities engaging in an aspect of State administration (*penyelenggaraan negara*) that receive government funding and non-government organisations funded by either the community or foreign sources (Articles 1(3) and 16). Specifically mentioned as public bodies are national and regional executive governments and legislatures, the judiciary and political parties (Articles 1(3) and 15).

The obligations that the FOI Law imposes on public bodies are significant. They must develop information and documentation systems to efficiently manage public information (Article 7(3)); create request-processing systems; and appoint employees to respond to requests (Article 13(1)(b)). The Law requires public bodies to publish six-monthly reports on their activities, performance and finances (Article 9(2)) and to proactively disclose information that could (*dapat*) 'threaten the necessities of life of the people and public order' (Article 10(1)). The FOI Law also requires that public bodies be ready to provide various types of public information 'at any time'. These types of information include: lists of the public information under their control; regulations and decisions they produce, along with the reasons for making them; policies and supporting documentation; and contracts with third parties (Article 11(1)). Public bodies must also report the number of information requests they receive, grant and reject each year and the time taken to fulfil these requests (Article 12). The Law also specifies particular information that must be disclosed by state-owned enterprises, political parties and NGOs (Articles 14–16).

To meet these requirements, Government Regulation 61 of 2010 stipulated that an information officer must be appointed in every public body by 23 August 2011 (Article 21(1)).[32] The main responsibilities of these officers are to receive and service requests for information; provide, store and protect information; conduct consequence assessments and classify information (discussed later, in 'Exemptions and exclusions'); determine whether embargoes

---

32 In the interim, these tasks were permitted to be performed by an agency's public relations or communications unit (Article 21).

on information should be lifted; and provide written explanations for policies made by the public body.[33]

Dispute resolution

The FOI Law also establishes mechanisms that information seekers can employ if public bodies ignore their requests or do not provide all of the information requested. Perhaps the most important of these mechanisms are provided by the central Information Commission, an independent institution which operates alongside a number of provincial information commissions.[34]

The FOI Law's provisions on requesting information and resolving disputes are as follows: Applicants must submit a written or oral request with the public body thought to possess the information (Article 22(1)). The public body must then respond in writing within 10 working days (Article 22(7)),[35] specifying whether the body possesses the requested information and, if so, whether the body is prepared to disclose it (Article 22(7)). The written response must include any proposed redactions (Article 22(7)(e)) and an estimation of costs to be borne by the applicant (Articles 22(7)(c) and (g)).

If the public body rejects the request, it must provide written reasons (Article 22(7)). The applicant has 30 days to lodge a formal objection with the information officer's superior if, under Articles 35(1) and (2), their application was refused or ignored; the information provided was not the information requested or was not provided within 10 days; or the estimated costs were excessive. The superior then has 30 days to issue a written response (Article 35(5)).

If still dissatisfied, the applicant has 14 days to apply for dispute resolution (Article 36(2)). In most cases, the first stage is voluntary mediation before the Information Commission, which must be completed within 100 working days (Article 38(2)). If both parties agree to the mediated outcome, the Commission issues a declaration containing the agreement reached. Once issued, this declaration becomes final and binding on the parties (Article 39).

On the other hand, if one or both parties are dissatisfied with the mediation, then the Commission can commence 'non-litigation adjudication' (*ajudikasi nonlitigasi*) (Article 42). Under this process, which appears to be very similar to arbitration, three Commissioners hear the dispute. These proceedings will usually be public unless they involve potentially excluded

---

33 In accordance with Government Regulation 61 of 2010 on the Implementation of the Freedom of Information Law, Article 14(1)(h).

34 The central Commission has seven members, with representatives from the government and community. They are chosen by the national parliament from a list compiled by the President (Articles 31(1) and (2)). Provincial commissions have five members (Articles 25(2) and (3)). Commission members elect their own chairperson and deputy chairperson (Article 25(4)).

35 However, the response time may be extended by a further seven working days, provided written reasons are given (Article 22(8)).

information (Article 43). The Commission has power to call applicants and officials to attend hearings and to request notes or materials possessed by the public body that are relevant to the disputed rejection (Article 27(1)). After hearing from both sides, the Commission can either uphold the rejection or order the public body to provide all or part of the requested information (Article 46(1)). Importantly, the public body bears the burden of showing that the information requested should not be disclosed, such as if the information is excluded by Article 17.

If either party is dissatisfied with the Information Commission's decision, then the FOI Law provides judicial avenues to resolve the dispute (Article 4(4)). Indonesia's administrative courts hear information disputes involving 'state public bodies' (*badan publik negara*) (Articles 47(1) and (2)) and the general courts hear cases involving other types of public body. These courts can order the public body to disclose all or part of the information or confirm the body's refusal to disclose (Articles 49(1) and (2)).

Exemptions and exclusions

The FOI Law declares that it seeks to limit the types of information that public bodies can or must keep secret (Article 2(2)). Most excluded information is listed in Chapter 5 of the Law, comprising Articles 17–20. Excluded is information that will:

- impede law enforcement (Article 17(a)(1))[36]
- threaten intellectual property rights (Article 17(a)(2)), national security,[37] national economic stability[38] or international relations[39]
- disclose Indonesia's natural resources (Article 17(d)), the contents of a private deed or will, or the personal information of an individual.

The FOI Law also excludes documents circulated within a public body (or sent between public bodies) that are, by their nature, confidential, unless the Information Commission or a court determines otherwise (Articles 17(i) and 20(a)).

---

36 Excluded is information about criminal investigations; the protection of witnesses, victims and law enforcement officials; and intelligence data related to preventing or handling transnational crime: Articles 17(a)(1)–(5).

37 Excluded information includes strategies, intelligence, operational details, tactics and techniques related to the defence and security of the nation; and the composition and disposition of force and capacity in defence and security.

38 Such as plans for buying or selling national foreign currency, shares, property and vital State assets; planned foreign investment and investigations into banking, insurance or other financial institutions: Article 17(e).

39 Such as the position, bargaining power and strategies that will be (or have been) used by States in international negotiations; correspondence, communication and code systems used in international relations; and the protection and security of Indonesia's strategic infrastructure: Article 17(f).

To classify information as exempt under these provisions, however, information officers in public bodies must first perform a careful and accurate 'consequences assessment' (Government Regulation 61 of 2010, Article 3(1)). The exemption must be in writing, specify the category of excluded information into which the information falls and include the reason for the exemption (Article 4(2)).

The Forestry Ministry and freedom of information

The success of any policy such as REDD+, which requires implementation by the Forestry Ministry, will depend to a large extent on the performance of the Ministry to be assessed, and this can only occur if there is access to information from the Ministry. In February 2011, the Forestry Minister issued a Regulation to implement the FOI Law in the Forestry Ministry.[40] The Regulation adopts many of the FOI Law's definitions, including of 'information' and 'public information'. It also incorporates key FOI Law principles. For example, it reconfirms the assumption that 'all public information is open and can be accessed by all users of public information, unless that information is exempt' (Article 2(1) of the Forestry Regulation). The Regulation also restates that the categories of exempt information are limited (Article 2(2)) and that the Ministry must provide requested information 'quickly' and 'simply' (Article 4(1)).

The Regulation, like the FOI Law itself, categorises information as needing to be proactively and periodically disclosed, available at all times, or produced on request (Article 6). General information, forestry statistics, audited financial and asset statements, performance accountability reports and action plans must be released periodically (Article 7); while information regarding hotspot distribution, forest fires and natural disasters – including volcanic eruptions, floods, landslides and wildlife disruption in forest areas – must be made accessible as soon as they are produced (Article 9).

The Regulation also sets out a long list of the types of public information that must be available at all times. This includes information about:

- forest plans and policies, including long-term development, national work, strategic and macro forestry plans
- forest and irrigation conservation areas, including information about their scope and spread, the status of forest gazettal processes, as well as forest allocation and function changes
- forest closures, changes to forest closures, socioeconomic conditions of communities who live around forests, and forest exploitation and use
- the names of those who hold various rights over forest areas and the territorial scope of those rights – including commercial rights over forest product exploration (*pemanfaatan hasil hutan kayu*, or IUPHHK); natural

---

40 Forestry Minister Regulation P.7/Menhut-II/2011 on Public Information Services in the Forestry Ministry.

forest (*hutan alam*, or HA); plantation forest (*hutan tanaman*, or HT); forest ecosystem restoration (*restorasi ekosistem*, or RE); non-timber forest product utilisation licence (*Izin Usaha Pemanfaatan Hasil Hutan Bukan Kayu*, or IUPHHBK); community plantation forest (*hutan tanaman rakyat*, or HTR); community forest (*hutan rakyat* (HR) or *hutan kemas-yarakatan* (HKM)); village forest (*hutan desa*, or HD); and reforested forest (*hutan tanaman hasil reboisasi*, or HTHR)

- use of forest areas, including for mining and infrastructure development
- annual harvesting quotas (*jatah penebangan tahunan*, or JPT)
- eco-tourism permit (*izin usaha wisata alam*)
- production and distribution of forest timber and non-timber products
- procedures to apply for permission to run forest enterprises
- production forests over which no rights have been granted, but over which exploitation permits are available
- provincial forest clearing data, impediments to use of forest areas, and forest nurseries
- rehabilitation of forests and land
- special-purpose forest areas
- conservation, biodiversity, tourism and environmental services areas and their scope, as well as the cultivation of bushland and wildlife
- forestry laws, planned and completed research and training, procurement of goods and services, forestry cooperation, international commitments, certification of conservation forest management and appointment of civil servants.

Exempt from disclosure are incomplete information or information for which intellectual property protection is being applied, confidential communications within the Ministry or within other public agencies, the results of internal audits and the personal data of Ministry officials (Article 10 of the Regulation). However, exempt information may be released with the approval of the Minister or the relevant official (Article 12).

The Regulation does not provide for the appointment of an information officer. Instead, all applications must be addressed to the Director of Public Relations (*Kepala Pusat Hubungan Masyarakat Kementerian Kehutanan*) at the Ministry's national, provincial, regional or district offices (Article 13(1)). Similarly, the Regulation does not establish a freedom of information division. All freedom of information requests are dealt with by the Ministry's Public Relations department.

The procedures for applying for information held by the Ministry of Forestry appear more onerous than those of the FOI Law. For example, the Regulation requires applicants to complete an information request form from the Ministry (Article 13(3)). It seems, then, that the Ministry will not accept simple requests 'in writing or otherwise', as recognised under the FOI Law (Article 22(1)). Further, unlike the FOI Law which requires the applicant to provide a reason for requesting the information, the Regulation requires that applicants disclose 'in detail their planned use of the requested information' (Article 13(5)). The ramifications of this extra requirement are potentially significant because it might provide additional legitimate grounds to reject requests for information, as discussed later.

Like the FOI Law, the Regulation requires the Ministry to respond to information requests within 10 days (Article 14(3)). Applicants have the right to receive reasons for rejection (Article 16(2)(c)). The Regulation also requires applicants to sign a document with the Forestry Ministry certifying that they will not use the information for illegal purposes (Article 16(1)(e)). If applicants breach this agreement, the Ministry has the right to make a claim against them, although the Regulation specifies neither the nature of that claim nor any penalties for breach (Article 17(2) (d)).

Implementation problems

Some commentators have criticised these FOI reforms. For example, not all of the required regional commissions have been established, and many government institutions have not yet appointed information officers (Saragih 2010; Saragih and Tampubolon 2010; Hukumonline 2012). Also, many public agencies have appeared to respond inadequately, or not at all, to information requests – some of which were lodged by NGOs to 'test' government compliance with the FOI Law.[41]

While many of these criticisms have some merit, they ignore the practical difficulties faced by public bodies in complying with the law, including that most have not received additional budgets to fund compliance. They also distract attention from much of the progress that has been achieved, particularly by the Information Commission in resolving information 'disputes'.

Information Commission decisions

The Information Commission has resolved dozens of information disputes, many of which appear on its website.[42] The Commission's approach in most cases is to categorise the information sought and then to determine if it is excluded under Article 17 of the FOI Law.[43] Many decisions emphasise that information held by public bodies is presumed to be 'public' and that it falls on public bodies to convince the Commission otherwise. If relevant, the

41 According to various reports, these uncovered various inadequacies, including a high proportion of rejected and ignored requests. See Basorie 2011; Freedom Info 2012; Erdianto et al. 2012; Thornley 2010.

42 See www.komisiinformasi.go.id.

43 See, for example, *Research and Application Discourse Institute* v. *PT Blora Patragas Hulu* (Information Commission Decision 001/VII/KIP-PS-A/2010).

Commission will also determine whether the information falls within any of the categories of information that public bodies must disclose.[44]

The Commission has rejected most of the arguments made by respondents attempting to defend their refusal to disclose requested information and has ordered the release of the information, sometimes with personal or sensitive information redacted.[45] For example, the Commission has consistently rejected claims made by public bodies that they lacked sufficient funds to find and provide the requested information, were unaware of the FOI Law or had no information officer.[46] The Commission has also rejected arguments put forward by the National Intelligence Agency that disclosure would prejudice national security[47] and by the police that law enforcement[48] would be compromised. In the vast majority of cases, the Commission has decided that the requested information is not excluded and has ordered the respondent to disclose all or some of it within 14 days, as required by Article 48(1) of the FOI Law.

Potential problems

The FOI Law contains a number of provisions that have the potential to result in large amounts of information being closed off. Problematic provisions include Articles 6 and 2(4).

Exclusion by internal regulation

Article 6 is entitled 'The Rights of Public Bodies' and declares, in paragraphs (1) and (2), that public bodies have the right to refuse to provide information that is 'excluded by' or 'does not accord' with 'written laws' (*peraturan*

---

44 See, for example, *Seknas Fitra v. National Mandate Party Central Executive* (Information Commission Decision 113/IV/KIP-PS-M-A/2011); *Muhammad HS v. Jakarta Provincial Government* (Information Commission Decision 63/II/KIP-PS-M-A/2010).

45 *LSM Sarvodaya v. Jakarta and Tangerang State-owned Electricity Companies* (Information Commission Decision 156/IV/KIP-PS-M-A/2012).

46 See, for example, *North Sumatra Indonesia Bible Institute v. Sunggal Senior High School 1 and Sunggal Junior High School 1* (Information Commission Decision 015/VIII/KIP-PS-M-A/2010); *Sarvodya v. General Directorate for Minerals and Coal, Energy and Natural Resources Ministry* (Information Commission Decision 178/V/KIP-PS-M-A/2012).

47 *Indonesia Corruption Watch Medan Branch v. Indonesia's State-owned Railway Company* (Information Commission Decision 298/VII/KIP-PS-M-A/2011) Although in *Fitra v. State Intelligence Agency* (Information Commission Decision 102/IV/KIP-PS-M-A/2011), the Commission ordered the disclosure of information that it classified as 'administrative', but that other types of information remain confidential. The Commission did not specify this information or justify why it should have been excluded, but it seems reasonable to speculate that this information related to 'national security', which, as mentioned, is excluded from disclosure under the FOI Law.

48 *Indonesia Corruption Watch v. Indonesia Police Headquarters* (Information Commission Decision 002/X/KIP-PS-A/2010).

perundang-undangan).[49] 'Written laws' are any form of government law, from statutes enacted by the national parliament through to government regulations, Presidential Instructions, Ministerial Decrees and Circulars, and even by-laws issued by local parliaments and regional heads – such as governors, regents and mayors. On a plain reading of Article 6(3), any such legal instrument could exclude almost any type of information. This interpretation would likely leave the FOI Law a dead letter. Many of these laws, such as instruments issued by Ministers and regional heads, only require the signature of a senior official to come into force. Public bodies seeking to avoid disclosure could, therefore, quite easily regulate to exclude particular information from disclosure requirements.

These concerns have not yet borne out in the Information Commission decisions, with the Commission avoiding the issue by refusing to interpret the 'other law' to require non-disclosure.[50] However, in a judicial appeal against a decision of the West Java Information Commission, the Bandung Administrative Court appeared to follow a National Audit Agency Regulation, which prohibited the disclosure of particular information,[51] even though the FOI Law appeared to require disclosure of similar, if not identical, documents.[52] In other words, the Court seems to have allowed public bodies to avoid disclosure by relying on internal regulations that define particular information as excluded. The danger here, of course, is that there seems to be nothing to prevent an agency from deliberately and unilaterally regulating to prevent disclosure of sensitive information.

The harm test

Article 2(4) establishes what some commentators have called the 'harm test' (Thornley 2010). The provision states that excluded information is information that is:

> [C]onfidential by reason of statute, appropriateness and the public interest, based on an assessment of the consequences that will arise if the information is disclosed to the community and after considering whether denying access to that information could protect a greater interest than the interest in opening access, or vice versa.

This provision appears to allow an information officer to refuse disclosure if he or she believes that disclosure of particular information will be

---

49 Article 6(3) then sets out types of information that public bodies must not disclose. These are categories also mentioned in Article 17: official secrets; and information that could endanger the State, individual rights or that protects industry from unfair competition.

50 *Busra Hasjim* v. *Kimia Farma Pension Fund* (Information Commission Decision 335/IX/KIP-PS-A/2011).

51 BPK Regulation 3 of 2011, Article 11(a).

52 *Depok Mayor* v. *Hidayat* (Bandung Administrative Court Decision 48/G/TUN/2012/PTUN-BDG).

more detrimental to the public interest than any advantage or benefit from disclosure. 'Appropriateness' and the 'public interest' are not defined in the FOI Law and are inherently vague and subjective. Their scope is broad and could potentially apply to almost any category of information.

It also appears that, when weighing up benefit and harm, information officers might not be prohibited from considering the purpose for which the information will be used. Article 4(3) of the FOI Law requires that applicants specify the reasons for seeking information. Some commentators have been critical of this requirement, noting that most countries do not impose it and that it is likely to simply arm public bodies with another basis on which to reject the request.[53] However, there seems to be nothing to stop an information officer or public body desiring to prevent disclosure of sensitive information from simply asserting that disclosure, for the purposes provided by the applicant, is not in the public interest. The statute and implementing regulations do not specifically permit the Commission or a court to review the merits of an information officer's decision to refuse disclosure.

In the decisions studied, the Information Commission has not yet performed a consequences assessment. Rather, the Commission's main concern appeared to be checking whether the public body itself had performed the assessment. Importantly, given the presumption that information is 'public', the Commission has emphasised that the onus falls on respondents to prove that they have performed a harm assessment. If they cannot, then the Commission will usually classify the requested information as public and order its disclosure. In most cases, the Commission found that an assessment was not made as required by the FOI Law.[54] Whether the Commission will evaluate the merits of a harm assessment that a government agency *has* performed remains an open question.

Conclusion

On paper, Indonesia's judicial system appears adequate to provide effective relief in many REDD-related cases. Traditional communities can bring claims against the government in administrative courts for 'administrative decisions' that violate their legal rights. Furthermore, they can seek compensation for loss in Indonesia's general courts under Article 1365 of the Civil Code. Article 1365 offers great flexibility, allowing the community to proceed even if they cannot prove that the defendant has breached a written law. They may also be able to pool their resources by bringing these claims as class actions. If unsatisfied with a decision issued by the lower courts, rights of appeal exist up to the Indonesian Supreme Court, which can even re-open

53 As Thornley (2010) puts it, 'public information ought to be public regardless of what the intended use is.'
54 See, for example, *Herunarsono* v. *International Rawamangun Primary School 12* (Information Commission Decision 002/II/KIP/PS-M-A/2011).

cases. Complaints against government and judicial decision making can be made to the National Ombudsman, which can then perform investigations and make recommendations to the government.

While structural reforms to Indonesia's judicial system make it unlikely that the courts will simply side with the government in these disputes, as they would likely have done during the Soeharto period, significant obstacles remain in the way of traditional communities seeking to challenge government action. Judicial corruption is perhaps the most significant of these. Commercial enterprises that hold lucrative licences or concessions are more likely than traditional communities to access funds to engage in bribery. They may seek to 'intervene' in disputes between citizens and the government agency official that awarded the licence or concession. For example, they might offer a bribe to ensure that courts uphold the validity of the licence. It seems there is little incentive for poorly paid judges to refuse such an offer, particularly if their impropriety is unlikely to be detected or pursued. Even if a traditional community were to obtain a favourable court judgment, it is likely to face significant difficulties in having that judgment enforced.

Since 2012 the Information Commission has heard and decided a substantial number of cases in which it has readily ordered disclosure. The main threat to an effective freedom of information regime appears to be the courts that hear appeals against Commission decisions. At time of writing, only eight appeals against Information Commission decisions had been posted on the Supreme Court's website,[55] which contains the most complete database of Indonesian court decisions.

In the majority of these cases, the administrative courts have not disturbed the Commission's decision.[56] However, a handful of Information Commission decisions in which disclosure had been ordered have been overturned. Yet an assessment of the likely future approach of the Indonesian courts in these cases is premature. As of early 2013 the Supreme Court had not yet published any appeals against administrative court decisions in information cases. It is to the Supreme Court, rather than first instance administrative courts, that the courts (perhaps even the Information Commission itself) are likely to look for guidance in future cases.

On balance it seems likely that the public, including communities affected by REDD+ projects, will be likely to successfully obtain relevant information relating to those projects, particularly from the Forestry Ministry. As mentioned, this includes information about: forest plans and policies; various types of licence, who holds them and over which land; use of forest areas, including for mining and infrastructure development; as well as forest statistics, financial statements, performance reports and action plans. The Forestry Minister has, by his own Regulation, declared that that the Ministry must

---

55 See putusan.mahkamahagung.go.id.
56 *Bangkalan Corruption Watch* v. *Bangkalan Parliament* (Surabaya Administrative Court Decision 75/G/2012/PTUN-SBY).

disclose these various types of information, provided that the application for information meets quite easily satisfied requirements stipulated in the Regulation.

If the Forestry Ministry rejects an information request, an appeal to the Information Commission is likely to yield an order for the Ministry to provide that information. Going by past practice, an administrative court or the Supreme Court will often, but not always, uphold that order if the Ministry decides to appeal.

One question of particular importance to REDD-related information requests remains unanswered the nature and scope of the 'natural resources information' exception under Article 17(d). What will constitute 'revealing natural resources'? Will information disclosing the previously unknown location of natural resources within a particular territory be excluded? Or will precise estimates of natural resource reserves be required? To our knowledge, neither the central nor regional information commissions have yet directly considered this exception, let alone these questions. And neither have the courts. However, if previous decisions of the Information Commission in which the Commission has interpreted the categories of excluded information are any guide, the relevant information commission would likely order disclosure in the absence of a compelling reason to maintain confidentiality. However, whether the administrative courts will choose to follow the commissions' decisions is by no means assured.

References

Agustina, R., H. N. Suharnoko and J. Hijma (2012) *Hukum perikatan*, Jakarta: Pustakan Larasan.

Asia Foundation and ACNielsen Indonesia (2001) *Survey report on citizens' perceptions of the Indonesian justice sector: Preliminary findings and recommendations*, Jakarta.

Basorie, W. D. (2011) 'Indonesia's freedom of information laws, one year on', *Jakarta Post*, 28 April.

Bedner, A. (2001) *Administrative courts in Indonesia: A socio-legal study*, Boston, MA: Kluwer Law International.

BPHN (National Legal Development Institute Team) (1993/94), *Naskah akademis peraturan perundang-undangan tentang perbuatan melawan hukum*, Jakarta.

Butt, S. (2007) 'Judicial review in Indonesia: Between civil law and accountability? A study of Constitutional Court decisions 2003–2005', PhD thesis, Law Faculty, Melbourne University.

Butt, S. (2008) 'Surat sakti: The decline of the authority of judicial decisions in Indonesia', in T. Lindsey (ed.) *Indonesia: Law and society*, 2nd edn, Annandale: Federation Press.

Butt, S. (2009) '"Unlawfulness" and corruption under Indonesian law', *Bulletin of Indonesian Economic Studies*, 45(2): 179.

Butt, S. (2012a) *Corruption and law in Indonesia*, London: Routledge.

Butt, S. (2012b) 'Indonesian Supreme Court invalidates its own ethics code', *East Asia Forum*, 1 March.

Butt, S. and T. Lindsey (2010a) 'Judicial Mafia: The courts and state illegality in Indonesia', in G. V. Klinken and E. Aspinall (eds) *The state and illegality in Indonesia*, Leiden: KITLV Press.

Butt, S. and T. Lindsey (2010b) 'Who owns the economy? Property rights, privatization, and the Indonesian Constitution: The Electricity Law Case', in A. McHarg, B. Barton and B. Bradbrook, *Property and the law in energy and natural resources*, Oxford: Oxford University Press.

Butt, S. and T. Lindsey (2012) *The Indonesian Constitution: A contextual analysis*, Oxford: Hart Publishing.

Crouch, M. (2007) 'The Yogyakarta Local Ombudsman: Promoting good governance through local support', *Asian Journal of Comparative Law*, 2(1): 1.

Crouch, M. (2008) 'Indonesia's national and local Ombudsman reforms: Salvaging a failed experiment?', in T. Lindsey (ed.) *Indonesia: Law and society*, 2nd edition, Annandale: Federation Press.

Erdianto, K., D. Aryani and M. Karanicolas (2012) *Implementation of the right to freedom of information: An assessment of three Indonesian public authorities*, Jakarta: Centre of Law and Democracy and Yayasan 28: www.law-democracy.org/wp-content/uploads/2010/07/Buku-UU-KIP-eng1.pdf.

Freedom Info (2012) 'Problems found in handling of RTI requests in Indonesia', freedom.org, 4 May 2012: www.freedominfo.org/2012/05/problems-found-in-handling-of-rti-requests-in-indonesia/.

Fuady, M. (2002) *Perbuatan melawan hukum (pendekatan kontemporer)*, Bandung: PT Citra Aditya Bakti.

Hukumonline (2012) 'UU keterbukaan informasi diabaikan', *Hukumonline*, 26 May.

Indonesian Corruption Watch (ICW) (2001) 'Menyingkap tabir mafia peradilan' (hasil monitoring peradilan).

Inside Indonesia (1997) 'Courting corruption', 49: www.insideindonesia.org/weekly-articles/courting-corruption.

Ismail, C. (2005) *Direksi dan komisiaris dalam perbuatan melawan hukum oleh perseroan terbatas: konstruksi hukum, tanggungjawab dan perlindungan hukum pihak ketiga*, Jakarta: Merlyn Press.

Legal Mafia Eradication Taskforce (2010) *Mafia Hukum*, Satgas Pemberantasan Mafia Hukum with the support of UNDP.

Mahkamah Agung (2012) *Laporan tahunan 2011*, Jakarta.

Mahkamah Agung (2013) *Laporan tahunan 2012*, Jakarta.

Ombudsman Repulik Indonesia (2010) *Laporan tahunan 2011*, Jakarta.

Pemberton, J. (1999) 'Open secrets: Excerpts from conversations with a Javanese lawyer, and a comment', in V. Rafael and R. Mrazek (eds) *Figures of criminality in Indonesia, the Philippines, and colonial Vietnam*, Ithaca, NY: Southeast Asia Program Publications, Cornell University.

Pompe, S. (2005) *The Indonesian Supreme Court: A study of institutional collapse*, Ithaca, NY: Southeast Asia Program, Cornell University.

Santosa, M. A., J. Khatarina and R. S. Assegaf (2012) 'Indonesia', in R. Lord, S. Goldberg, L. Rajamani and J. Brunnee (eds) *Climate change liability*, Cambridge: Cambridge University Press.

Saragih, B. B. T. (2010) 'Red tape hinders access to information', *Jakarta Post*, 1 May.

Saragih, B. B. T. and H. D. Tampubolon (2010) 'Access to info improved despite poor preparations', *Jakarta Post*, 1 May.

Sherlock, S. (2002) 'Combating corruption in Indonesia? The Ombudsman and the Assets Auditing Commission', *Bulletin of Indonesian Economic Studies*, 38(3): 367.

Tasrif, S. (1971) *Menegakkan rule of law dibawah Orde Baru*, vol. 1, Jakarta: Persatuan Advokat Indonesia (PERADIN).

Tempo Interaktif (1997) 'Wawancara Adi Andojo: "Akhirnya karir saya dan juga Mahkamah Agung tidak ternoda"', *Tempo Interaktif*.

Thornley, A. (2010) 'We have a right to know. Is our new law helping us find out?', *Jakarta Post*, 26 December.

World Bank (2003) *Combating corruption in Indonesia: Enhancing accountability for development*, Jakarta: East Asia Poverty Reduction and Economic Management Unit.

World Bank (2004) *Village justice in Indonesia: Case studies on access to justice, village democracy and governance*, Jakarta.

9 The Indonesian Constitutional Court and REDD+

The Indonesian Constitutional Court (the Court) was established in 2003. It has developed into Indonesia's most respected judicial institution and one of the most successful products of the reformation (*Reformasi*) movement that emerged in Indonesia when Soeharto fell in 1998 after 33 years in power. Under Soeharto's authoritarian regime, the judiciary, dependent on the government and lacking judicial review powers, had been largely reduced to a rubberstamp for government action and standards of competence and integrity had plummeted. Defying the expectations of many, the Constitutional Court emerged as a professional and respected judicial institution. In the decade since its establishment, the Constitutional Court has drawn much praise for striking down many statutory provisions that it found to be in breach of Indonesia's Bill of Rights, inserted into the Indonesian Constitution in 2000 as part of the post-Soeharto reforms.

The Court has, however, issued decisions in recent years that are likely to have significant ramifications for any national statutory scheme on REDD+. The national legislature must, we argue, pay heed to these decisions when formulating REDD+ legislation. If not, the legislature risks having parts of any statute invalidated by the Court, as it has in cases involving natural resources legislation. If past cases are any guide, Indonesian customary law communities, probably represented by capable non-government organisations (NGOs) such as the Indigenous People's Alliance of Archipelago (*Aliansi Masyarakat Adat Nusantara*, or AMAN), will inevitably challenge any problematic aspects of the statute before the Constitutional Court, claiming a breach of constitutional principles.

From a reading of these cases, we argue that the Court is likely to uphold any such constitutional challenge, to the extent that it employs at least one of the three main constitutional grounds that the Court has accepted in these previous cases. First, the Court has, in a string of cases, held that the State must respect, protect and fulfil the rights of traditional communities when seeking to regulate natural resources, particularly in exploration and extraction. The Court is, therefore, likely to require respect and protection for all pre-existing rights and entitlements of customary law communities to the forests in which REDD+ is proposed. The Court has also granted customary law communities some form of entitlement to participate in decision making about specific natural resources, particularly if their access to those resources is at stake.

Second, the Court would likely strike down national parliamentary attempts to reserve for national government institutions the power to unilaterally designate forests for REDD+ or to claim control over the use (or non-use) of forest resources. The Court will probably require that local governments have a 'real' say in how local natural resources are used or 'fenced off' as REDD+ seems to require. The national government will probably be unable to 'impose' REDD+ in the face of resistance from local governments.

Third, the Court has consistently required that the State maintain a relatively high level of control over natural resources. This has presented constitutional impediments to the government giving itself power, at least by statute, to grant a licence to another party to control or otherwise manage natural resources, thereby relinquishing its own control over those resources. To the extent that REDD+ requires the State to abstain from exercising this control, the Court is likely to find it unconstitutional. Going by previous cases, the Court might even decide that the State cannot commit to 'fence off' land for long periods, on the basis that doing so would be tantamount to relinquishing State control over natural resources.

The Constitutional Court: jurisdiction, precedent and decision making

The Court's jurisdiction is defined in Article 24C of the Indonesian Constitution and in Article 10 of the statute establishing the Court: Law 24 of 2003 on the Constitutional Court. Its primary function is judicial or constitutional review; however, it has several other additional functions, including resolving disputes between State institutions established by the Constitution about their relative jurisdictions, about the dissolution of political parties and about general election results. The Court must also 'provide a decision' if the national parliament refers to it a suspicion that the President or Vice President has committed an act of treason, corruption or bribery; another type of serious crime or improper conduct; or no longer fulfils the constitutional requirements to hold office.

The Court's constitutional review jurisdiction is limited to reviewing national statutes against the Constitution. This is a significant restriction and the Court cannot, therefore, assess the constitutionality of lower order laws, such as Ministerial Regulations and Decisions, including those that purport to specifically regulate aspects of REDD+.[1] However, this does not mean that the Court cannot or will not be able to influence REDD-related

---

1 This chapter draws on Butt (2014). Only the Indonesian Supreme Court has the power to review these types of lower order law, but only against statutes. We discuss this issue in Chapter 7.

laws. Provisions relevant to REDD+ can be found in existing statutes such as Law 41 of 1999 on Forestry (the Forestry Law) – which is legislation enacted by the national parliament. Also, if the national parliament issues a specific statute to regulate REDD+, then that statute will fall within the Court's constitutional review jurisdiction.

Formally, the Constitutional Court is not bound to follow its own previous decisions. As a civil law jurisdiction, Indonesia does not have a formal system of precedent and there are no provisions in the Constitution, or in the statute governing the Constitutional Court, that require the Court to issue consistent decisions. However, as we shall see, the Court generally does follow and quote from its own previous decisions. The Court is also clearly willing to extend its reasoning in cases involving particular natural resources or important industries – such as electricity, mining and waterways – to other resources or industries, such as land, forests and petroleum. This is important because even though the Court has decided several cases about forestry, most of its cases about natural resources relate to other types of resource. Further, the Court has only heard cases about the 'exploitation' of natural resources, not their 'preservation' as REDD+ seeks to achieve. Nevertheless, judging by past practice, the Court is likely to extend key aspects of these previous decisions to a statutory REDD+ scheme.

The Court has, generally speaking, given its decisions only prospective effect. In other words, if the Court decides that a statute breaches the Constitution and declares it invalid, that statute will only be invalid from the moment the Court finishes reading its decision. Any action taken under the statute between its enactment and its invalidation is not affected by the declaration of invalidity and therefore remains legal. The Court has emphasised the prospectivity of its decisions in several natural resource cases (discussed later).[2] Any contracts or concessions made or issued on the basis of a provision that the Court later declares invalid will remain valid until their expiry. In this context, a Constitutional Court decision would only prevent application of the invalidated provision after the Court had read its decision. If a constitutional challenge to a national statute that sought to provide the legal infrastructure for REDD+ projects were successful, the decision could not be used to shut down any projects that had already been set up under this statute later declared unconstitutional by the Court. The decision would only render invalid any future projects set up under the statutory provisions that the Constitutional Court had struck down.

The Court quite commonly issues decisions in which it declares statutory provisions to be either 'conditionally constitutional' or 'conditionally unconstitutional'. The effect of these decisions is that the statute under review will

---

2 For example, Constitutional Court Decision 3/PUU-VII/2010, reviewing Law 27 of 2007 on the Management of Coastal Areas and Small Islands (*Coastal and Remote Areas Law* case (2010)) [3.15.13]; Constitutional Court Decision 36/PUU-X/2012, reviewing Law 22 of 2001 on Oil and Natural Gas (*Oil and Natural Gas Law* case (2012)) [3.21].

be invalid unless interpreted in line with conditions imposed by the Court. In its early years, the Court's conditions were relatively vague. For example, in its first 'conditionally constitutional' decision, the Court upheld the constitutionality of a statute allowing film censorship, but said that it needed to be interpreted in line with the 'spirit of the times' – that is, the 'spirit of democracy' and 'respect for human rights'.[3] More recently, however, the Court has been far more specific in its conditions so that they now resemble direct and precise legislative amendments. Some of the decisions discussed in this chapter fall into this category. However, the decisions with these amending effects are not cast by the Court as legislative amendments. The Court continues to proclaim itself as a negative legislator (that is, it can only strike down legislation) and denies being a positive legislator (that is, it can make law).[4]

Maintaining State control

Article 33(3) of the Constitution states: 'The earth and water and the natural resources contained within them are to be controlled by the state and used for the greatest possible prosperity of the people.' The Constitutional Court has invalidated many statutory provisions – and even an entire statute – for breaching Article 33(3), based on what some commentators would describe as a conservative interpretation of what the State must do to retain its 'control' over natural resources (Butt and Lindsey 2008, 2012; Afghani 2013).

The Court first considered the meaning and requirements of 'state control' in Article 33 in the *Electricity Law* case.[5] Before this Law was enacted, there had been little competition in Indonesia's electricity market, with state-owned enterprises (SOEs) holding a virtual monopoly over activities in the sector. The Electricity Law purported to 'unbundle' the provision of electricity into various activities – including generation, distribution and sale – and sought to allow significant private sector involvement in those activities. The main question facing the Court was whether this unbundling breached Article 33(2) of the Constitution. That provision is similar to Article 33(3), except that it requires the State to control important branches of industry. In coming to its decision, the Court defined 'state control' and established the following principles to which it has consistently referred in subsequent cases.

---

3 Constitutional Court Decision31/PUU-IV/2006, reviewing Law 32 of 2002 on Broadcasting (*Broadcasting Law* case (2006)).
4 The Court has begun issuing more of these types of decisions since the Chief Justiceship of Mahfud M. D. (2008–13). Provisions were invalidated in 43 of the 157 cases heard during the five-year term of Mahfud's predecessor, Professor Jimly Asshiddiqie (2003–08). Of these 43 decisions, 15 included declarations of conditional constitutionality – around 35 per cent. During Mahfud's term, almost 60 per cent of successful review applications involved declarations of conditional constitutionality – that is, in 50 of the 84 cases in which the court upheld constitutional challenges, out of a total of 303 lodged cases.
5 Constitutional Court Decision 001-021-022/PUU-I/2003, reviewing Law 20 of 2002 on Electricity (*Electricity Law* case (2003)).

The State's power to regulate natural resources did not, of itself, constitute State control because the State already had an inherent power to regulate, irrespective of Article 33. Furthermore, mere civil ownership by the State was not 'control by the state' because natural resources were public assets collectively owned by all Indonesians and the State was required under Article 33 to control those assets for the greatest possible collective prosperity. Rather, State control comprised five activities: policymaking, administration, regulation, management and supervision. In addition, at least for natural resources, these five activities must be performed for one purpose: the greatest prosperity of the people. The Court gave examples of activities that would constitute these elements of State control: the government could exercise its power to administer by issuing and revoking licences and concessions; it could manage the relevant natural resources sector or industry through share ownership or by running the enterprise as a State institution.

The Court found that the Electricity Law sought to privatise aspects of the electricity industry and, therefore, relinquished State control to such a degree that it breached Article 33(2). Rather than merely invalidating the provisions that unbundled the sector, the Court invalidated the entire statute, deciding the Law's main thrust was privatisation of an important sector, which Article 33(2) prohibited. The Court has since reviewed statutes dealing with many of the natural resources that once were, or still are, plentiful in Indonesia – including land, forests, water, oil and natural gas, coastal areas, minerals, coal – and invalidated or declared various provisions to be 'conditionally constitutional' (Butt and Lindsey 2012). In some of those cases, the Court appears to have imposed even stricter obligations on the State to meet its 'control' requirements under Article 33(3). We now turn to discuss three recent cases in which it has done so.

Investment Law *case (2007)*[6]

In 2007, Indonesia's national parliament enacted an Investment Law aimed at improving Indonesia's climate for both foreign and domestic investment.[7] The Law sought to address an issue that foreign investors commonly complain about – the difficulty in obtaining rights over land, at least for an extended period of time. The Law aimed to make it possible for foreigners to hold various land rights for several decades – including the right to cultivate (*hak guna usaha*), build (*hak guna bangunan*) and use (*hak pakai*) land.[8]

The pre-existing law had allowed these rights to be extended, at the absolute discretion of the government, for 20 to 30 years (depending on the right)

---

6 Constitutional Court Decision 21-22/PUU-V/2007, reviewing Law 25 of 2007 on Investment (*Investment Law Case* (2007)). This description draws on Butt and Lindsey (2012).
7 Law 25 of 2007 on Investment (which replaced Law 1 of 1967 on Foreign Investment and Law 6 of 1968 on Investment).
8 Foreigners remain unable to obtain freehold title (*hak milik*).

after expiry of the initial grant of between 25 and 35 years. The 2007 Law purported to allow investors to extend these rights 'upfront' – that is, at the time of their initial grant. Investors would qualify for this 'inducement' if their investment improved the competitiveness of Indonesia's economy, involved high risk and did not damage the public interest (Article 22). Article 22(4) allowed the government to revoke the grant if the recipient neglected the land, used the land contrary to the purposes of its grant or otherwise damaged the public interest. In the *Investment Law* case (2007), the Indonesian Chamber of Commerce and several individual applicants challenged the constitutionality of these upfront extensions, including on the basis that they breached Article 33(3).

After confirming that land fell within Article 33(3) and therefore the State was required to 'control' it, the Constitutional Court turned to whether the State relinquished control if it renewed or extended land rights after they had expired. The Court decided that it did not. The State retained power to 'regulate' the land because the State could determine the term of that right and could attach conditions to it. Supervisory control was maintained because the State could impose sanctions for misuse of the land.

However, the upfront extensions were another matter. The Court held that they were unconstitutional because they had the potential to 'reduce or remove' the State control required by Article 33(3) (*Investment Law* case 2007: 263). The Court seemed concerned that upfront extensions left the State without the absolute discretion to revoke or refuse to extend these rights – a discretion that it had enjoyed under the pre-existing law. In this context, the Court found that Article 22(4) of the Investment Law, which limited the State's power to revoke the grant to particular circumstances, impeded – and perhaps even relinquished – the State's right of absolute control (*Investment Law* case 2007: 258). The Court also emphasised Article 32 of the Investment Law, which allowed foreign investors to arbitrate disputes with the government before international forums. Such disputes could arise if Article 22(4) were applied against an investor. If the State were to lose such arbitrations, its 'control' would be lost – at least for the remainder of the term of the land right. According to the Court:

> The interest to be protected by [Article 33(3)] of the Constitution is the welfare of the people in relation to the exploitation of the earth, water and resources contained therein. In relation to land, this interest was translated into national land policy intended to achieve the welfare of the people, including by redistributing ownership of land and limiting the permissible amount of owned land, so that land control and ownership is not concentrated in the hand of a group of people. This was among the things achieved through the Agrarian Law (Law 5 of 1960) and Law 56 of 1960 on Restriction of Agricultural Land Holdings. This restriction and distribution means that economic sources are spread more evenly and, ultimately, that the goals of people's prosperity will be achieved

equitably. Further, for land that is controlled by the State, the even distribution of land rights is achieved through a policy of equal opportunity to obtain various land rights ... for a limited and not excessive period. (*Investment Law* case 2007: 256)

This case is likely to affect the way that REDD+ can be legally structured in Indonesia. In particular, statutory REDD+ schemes that require land rights to be held directly by investors or intermediaries for long periods are likely to be unconstitutional, unless the State maintains its control by granting itself unbridled power to revoke those rights. This is, of course, unlikely to be attractive to investors.

Coastal and Remote Areas Law *case (2010)*

The 2007 Coastal Management Law (the 2007 Law)[9] classified coastal areas into various zones, including conservation areas, fishing areas, shipping lanes, port areas and public beaches. It also purported to authorise the government to grant coastal water concessions (*hak pengusahaan perairan pesisir*) over coastal waters for aquaculture and tourism to the private sector, including to foreigners (Articles 23(2), 23(6) and 23(7)). Concessions could extend over resources contained from the water surface to the ocean floor (Article 16) and could last up to 60 years, with an initial period of 20 years being twice extendable (Article 16). They could be transferred to another party or used as collateral (Article 20). To obtain a concession, applicants needed to meet various administrative, technical and operational requirements (Article 21).

In this case, various leaders of customary law communities located in the coastal areas challenged, with the assistance of NGOs, various provisions of the 2007 Law.[10] These communities had relied on coastal resources for their livelihoods and were concerned that these provisions would allow the government to award concessions that restricted their traditional rights to access and use these resources or, indeed, to close them off altogether.

The Constitutional Court accepted that, as a matter of fact, traditional communities, including those with which the applicants were associated, existed in many of the coastal areas governed by the Law. The Court also accepted that these communities followed customary laws that governed their use of natural resources. The Court decided that provisions of the Law that allowed the government to grant these concessions over coastal areas were constitutionally invalid on several grounds, some of which are discussed later.[11]

---

9 Law 27 of 2007 on the Management of Coastal Areas and Small Islands.
10 On 14 separate grounds (*Coastal and Remote Areas Law* case 2010: [3.9]). On these grounds, they challenged Articles 1(4), 1(7), 1(18), 14(1), 16(1), 18, 20, 21(1), 21(2), 21(3), 21(4), 21(5), 23(1), 23(2), 23(4), 23(5), 23(6) and 60(1).
11 Articles 1(18), 16, 17, 18, 19, 20, 21, 22, 23(4), 23(5), 50, 51, 60(1), 71 and 75 of the 1945 Constitution. It seems that some of these grounds overlap, even though they are contained in

As for the Article 33(3) ground, the Court found that coastal areas, and the natural resources within them, clearly fell within Article 33(3)'s ambit. The State was, therefore, required to exercise control over coastal area resources by formulating policies and issuing regulations and by administering, managing and supervising them. However, the Court found that the State had relinquished this control when it issued these concessions. This case raises questions about whether the national government could establish a REDD+ scheme by issuing concessions over large areas without relinquishing State control, as it did in the Coastal Management Law.

Oil and Natural Gas Law *case (2012)*[12]

Perhaps the most significant recent case in which the Constitutional Court has applied Article 33(3) to invalidate provisions of a statute is the *Oil and Natural Gas Law* case (2012). In this case, 10 Islamic organisations and 32 individual applicants asked the Court to review the constitutionality of Law 22 of 2001 on Oil and Natural Gas (the 2001 Law) on various grounds. The applicants claimed that by creating BP Migas – a State agency that the government had established to regulate and monitor the oil and gas sector – the 2001 Law reduced State control over natural resources, thereby breaching Article 33(3) of the Constitution. Under the 2001 Law, BP Migas' main functions were to enter into cooperation contracts and to monitor their implementation to ensure that oil and gas resources generated the maximum benefit for the greatest prosperity of the people (Articles 44(1) and (2)). BP Migas also advised the Energy and Mineral Resources Minister on cooperation contracts, production plans, budgets and the appointment of oil and gas sellers, again in the interests of securing the largest possible benefit for the people via the State (Article 44(3)).

The applicants argued that the contracts BP Migas had entered into with foreign companies, particularly for the exploration and extraction of petroleum, restricted the State's ability to freely regulate and control oil and gas resources, simply because those contracts bound the State. The applicants were also concerned that many of these contracts exposed the government to binding international arbitration (as required by Articles 1(23), 4(3) and 44)). For the applicants, not only would this impose a financial burden on the State; it also undermined both the national parliament's authority as the people's representatives and participation by the people as the owners of natural resources. Finally, the applicants complained that by allowing commercial enterprises to operate in the oil and gas sector (such as in Articles 3(b) and 9), the 2001 Law undermined State control, which required SOEs to compete with other operators.

---

separate provisions of the Constitution. Indeed, the Court dealt with some of these separate constitutional provisions together.

12 Description of this case draws from Butt and Siregar (2013).

An 8:1 majority of the Constitutional Court agreed with many of these arguments, deciding to excise from the 2001 Oil and Natural Gas Law all references to BP Migas, including the provisions granting it powers and functions (*Oil and Natural Gas Law* case 2012: [3.13.5]). By so doing, the court disbanded BP Migas, basing its decision on Article 33(3) of the Constitution. The majority categorised each of these five activities that comprise 'state control' – regulation, policymaking, management, administration and supervision – into one of three 'tiers' or levels of importance, depending on the extent to which the majority thought that activity achieved the greatest possible prosperity of the people.[13] Direct management over the natural resource was 'the most important first-order form of state control' (*Oil and Natural Gas Law* case 2012: [3.12]). According to the majority, direct State management of natural resources through SOEs ensured that all profits would flow to the State, thereby indirectly bringing greater benefits to the people. By contrast, handing over natural resources management to the private sector meant sharing profits between the State and private entities, thereby reducing the benefits flowing to the people. The majority decided that the State needed to fully manage natural resources unless it was unable to do so, in which case, opportunities could be given to foreigners. However, if the State had sufficient capital, technology and capacity to manage the natural resource, then the State must directly manage that natural resource. Of secondary importance were, equally, policymaking and administration. Both regulation and monitoring fell within the third tier.

Applying this interpretation to BP Migas, the majority decided that its statutory functions were insufficient to constitute 'control' of the sector as required by Article 33(3), largely because it did not 'directly manage' oil and gas resources. Upstream oil and gas activities were managed by the commercial entities with which BP Migas contracted – whether SOES (*badan usaha milik negara*, or BUMN), regional SOEs (*badan usaha milik daerah*, or BUMD), cooperatives or private enterprises – not BP Migas itself. The Court also decided that when BP Migas entered into a contract with a private enterprise, the prosperity of the people was not 'maximised' because the people would have to share any profits derived from natural resources with the private enterprise.

The Court also held that the contracts BP Migas made with commercial entities to engage in upstream activities undermined State control. According to the majority:

> Once the contract is signed, the government is bound by the contract. The government loses sovereignty and control over natural resources so

13 Although the court did not explain why it decided to rank the activities and how it devised the ranking. The rationale for ranking direct management as the most important aspect of State control and regulation as the equal least important is unclear, because it appears that 'regulation also [includes] supervisory activities, as well as license-granting, standard-setting, in addition to the traditional understanding of enacting rules' (Afghani 2013).

that exercising that control might breach the contract. However, as representatives of the people and the controller of natural resources, the state needs freedom to make regulations that bring the greatest possible prosperity to the people ... According to the Court, the relationship between the state and the private sector in the management of natural resources cannot be established through civil law. It is a public relationship ... [because it involves] providing concessions or licences that are under the complete control and power of the state. Civil contracts degrade the sovereignty of the nation over natural resources – in this case oil and natural gas ... To avoid this problem, the government can establish or appoint a state-owned enterprise and give it a concession to manage oil and natural gas in ... a Working Area so that that state-owned enterprise is the one entering into contracts with commercial enterprises. In this way, there is no longer a connection between the state and the commercial enterprise.[14] (*Oil and Natural Gas Law* case 2012: [3.13.3])

Until the government could issue new legislation in response to its decision, the Court declared that BP Migas' functions were to be performed by the Energy and Mineral Resources Ministry (*Oil and Natural Gas Law* case 2012: [3.22]).[15]

This decision appears to present very significant obstacles to REDD+. In particular, it seems to require the State to 'directly manage' any statutory REDD+ scheme, unless it lacks the capacity to do so. Unfortunately, the decision obscures the extent to which the private sector can legally participate in any such scheme, if at all, because the majority did not explain how the capacity of the State should be assessed. The majority appeared to reserve the power to determine State capacity for itself and, presumably, decided that the State was, in fact, capable of managing the oil and natural gas sector, without explicitly declaring that it was capable. Only Justice Harjono, in dissent, accepted that whether the State has sufficient capacity and capital to directly

14 In his sole dissent, Justice Harjono agreed that those contracts bound the State, but disagreed with the majority that any ensuing constraints on the State breached Article 33 of the Constitution for interfering with the State's 'control' of the natural resources to which the contract applied. Harjono emphasised that Indonesia is a 'law State' (*negara hukum*) and that the State could not simply use its power over national resources as it deemed fit, once it had entered into such a contract. Rather, for Harjono, the State control requirement was met because the State controlled BP Migas. Its chairperson was appointed and dismissed by the President, after consultation with the national parliament. According to Harjono, the State (through BP Migas) exercised control over the sector when it negotiated contracts and awarded concessions. After agreements had been made and contracts signed, the control had already been exercised and the Indonesian government was bound by the contract.

15 The decision did not mean, however, that the contracts which BP Migas had entered into before its disbandment were invalid. The Court declared that, in the interests of legal certainty, all working contracts made between BP Migas and commercial enterprises would continue in force until their expiry or a date on which the parties agreed (*Oil and Natural Gas Law* case 2012: [3.21]).

manage the sector was a matter for the President and the national parliament, who knew more about these matters than the Court.[16]

As mentioned, the majority appeared to fear the State signing away its absolute control through contracting with a private party. This seems similar to the Court's concern, expressed in the *Investment Law* case, about the State losing control by opening itself to binding international arbitration. The extent to which the State can directly enter into contracts with private sector enterprises to administer or operate REDD+ projects appears now to be unclear, at least to the extent that the contract binds the State to refrain from exercising ultimate and perhaps even unbridled control over natural resources.

Upholding traditional rights

Article 18B(2) of the Constitution declares that: 'The state recognises and respects customary law communities and their traditional rights provided they still exist and are in accordance with community developments and the principle of the unitary Republic of Indonesia as regulated by statute.' According to Article 28A, '[e]very person has the right to live and the right to maintain their life and livelihood.'

In a line of recent cases, the Constitutional Court has held that when the State awards concessions over resources and land that are relied upon – sometimes for generations – by recognised individuals or customary law communities, it breaches the constitutional rights of those communities to 'recognition, protection and fulfilment of their traditional rights' (Article 18B(2)). In a handful of cases, the Court has also found that awarding these concessions breaches Article 28A and the State's obligation in the Preamble to protect the people and their welfare. Relevantly, the Preamble provides that the purposes of establishing the State of Indonesia include 'to protect the entire nation ... to advance public prosperity, to enlighten the life of the nation, and to participate in realising world order based on independence, civilised peace, and social justice'. The Court has found that issuing concessions over traditional land can even breach Article 33(3). In some of these cases, the Court has also granted customary law communities an entitlement to participate in decision making about specific natural resources, particularly if their access to those resources is at stake, though it has not provided details about the nature of that entitlement.

---

16 The Court also did not consider any positive effects of non-government involvement on the oil and gas industry and, in particular, whether such involvement might create greater prosperity for the people than if the government directly managed it. As Justice Harjono pointed out, the oil and gas sector is high risk, requiring significant capital and capacity. Opening up competition and allowing private sector involvement in upstream activities might therefore allow more exploration and exploitation to take place, compared to what the State could achieve. This would result in profits or other benefits that, even if split between the industry participant and the State, might be significantly more than had the State directly managed the activities itself.

Perhaps the most significant of these cases, and the most relevant to REDD+, is the *Traditional Forest Community* case, handed down in May 2013, to which we now turn.

Traditional Forest Community *case (2012)*[17]

This case was brought by the AMAN[18] NGO and two customary law communities who claimed that the State had denied them their constitutional traditional forest rights.[19] They pointed to several instances in which the government had taken over customary forests and turned them into state-owned lands.[20] This, they claimed, had occurred when their traditional forest rights were categorised as 'State forest', which falls within the exclusive control of the State under provisions of Law 41 of 1999 on Forestry (the 1999 Forestry Law). The applicants argued that these provisions allowed the State to award rights over traditional forests to commercial entities without obtaining the agreement of the traditional community that used or occupied those forests and without being required to compensate its members. The result was that traditional communities were excluded from forestry resources that they had accessed for generations (*Traditional Forest Community* case 2012: [3.13.1]). The applicants asked the Constitutional Court to excise from the Forestry Law the provisions that defined State forests to encompass customary law forests and to 'reformulate' provisions that, they argued, otherwise breached their constitutional rights.

The Court agreed with the applicants' principal arguments and, by issuing declarations of conditional constitutionality, amended the 1999 Forestry Law to remove customary forests (*hutan adat*) from the definition of 'State forest'. More specifically, the Court decided that Article 1(6) of the Forestry Law would be invalid unless it were interpreted to remove the word 'State'. Customary forest was thus redefined as 'State forest located within a customary law community area'. The Court also imposed a condition on the constitutionality of Article 5(1), which states that forests 'consist of (a) state forest (*hutan negara*), and (b) forestry over which rights have been granted (*hutan hak*)'. The Court decided that Article 5(1) would be unconstitutional unless interpreted to add: 'State forest, referred to in (1)(a), does not include customary forest.'[21]

---

17 Constitutional Court Decision 35/PUU-X/2012, reviewing Law 41 of 1999 on Forestry (*Traditional Forest Community* case (2012)).
18 The word *aman* means 'safe' in Indonesian.
19 The Kuntu and Cisitu communities.
20 Including in Kasepuhan, Lebak, Banten in 1992 (Pasandaran 2013).
21 The Court also invalidated Article 5(2) and the Elucidation to Article 5(1) and amended Article 5(3), presumably because the amendments made by the Court to Article 5(1) had the effect of rendering obsolete the Elucidation to Article 5(2) and aspects of Article 5(3). Article 5(2), for example, had declared that: 'State forest, referred to in Article 1, can comprise customary forest.' Article 5(3) had stated that: 'The government determines the status of forests as referred to in

The Court also made several requested changes to Article 4(3) of the Forestry Law. This provision had originally read: 'The state is to observe the rights of customary law communities when exercising control over forests, provided that the community in fact exists and its existence is recognised, and that this does not the contradict the national interest.' The Court held that Article 4(3) would be unconstitutional and have no binding force unless it were given the following meaning: 'The state is to observe the rights of customary law communities when exercising control over forests, provided that they still exist and accord with community developments and the principle of the unitary Republic of Indonesia regulated by statute.'

The main legal arguments the Court employed to reach its decision were as follows. According to the Court, Article 18B(2) of the Constitution gave traditional communities the right to recognition and to have their traditional rights protected as constitutional rights (*Traditional Forest Community* case 2012: [3.12.1]).[22] The Court explained that when the Indonesian people decided to establish the State of Indonesia in 1945, they chose to establish a welfare state (*negara kesejahteraan*). This, the Court argued, was clear from the Preamble of the Constitution, mentioned earlier. This choice was also clear from the national ideology, *Pancasila*, particularly its inclusion of 'social justice' as one of its five pillars.[23] Therefore, the State needed to 'work hard' to achieve welfare for all people, including those who lived in traditional communities and relied on natural resources.

However, while the Constitution recognised traditional communities as 'right bearers' (*penyandang hak*) and legal subjects, the Forestry Law did not grant them the same rights as other legal subjects. In particular, the Forestry Law granted clear powers and rights to the State and to entities with state-issued rights over forests, but any entitlements of traditional communities to land and forests were unclear and tenuous (*Traditional Forest Community* case 2012: [3.12.2]). The Law's inclusion of customary law forest within State forest, and its failure to clearly provide traditional rights and entitlements,

Articles 5(1) and 5(2); and customary forest is to be stipulated provided that [it is] in accordance with the reality of the customary law community in question still exists and its existence is recognised.' The provision now reads: 'The government determines the status of forests as referred to in Articles 5(1) and 5(2); and customary forest is to be stipulated provided that [it is] in accordance with the reality of the customary law community in question still exists and its existence is recognised.'

22 The Court pointed to the many statutory reforms that recognised customary law communities, enacted since the Constitution was amended to include Article 18B(2): the Forestry Law, Law 39 of 1999 on Human Rights, Law 32 of 2004 on Regional Government, Law 31 of 2004 on Fisheries, and Law 27 of 2007 on the Management of Coastal Areas and Small Islands.

23 *Pancasila* contains the following principles:
  1. *ketuhanan yang maha esa* (belief in unitary deity)
  2. *kemanusiaan yang adil dan beradab* (a just and civilised humanity)
  3. *persatuan Indonesia* (the unity of Indonesia)
  4. *demokrasi* (democracy)
  5. *keadilan sosial* (social justice).

meant that in practice, traditional communities lost access to the forest resources on which many depended for their livelihoods, often resulting in conflict (*Traditional Forest Community* case 2012: [3.12.3], [3.13.1]). The result was injustice and legal uncertainty for traditional communities:

> Customary law communities occupy a weak position because their rights are not clearly and firmly recognised when up against the state with very strong control. State control over forests should in fact be used to allocate natural resources justly in the interests of the greatest possible prosperity of the people. (*Traditional Forest Community* case 2012: [3.12.4])

Ultimately for the Court, while the State had 'full authority to regulate and decide upon the availability, allocation, exploitation, administration of forests and legal relationships arising therein' in respect of customary law forest, its authority was:

> [L]imited by the customary law of the forest community. Traditional community forest (also referred to as kinship forest and sovereign forest, amongst others) is governed by *hak ulayat*, which exists within the territory of a single traditional community. Traditions are followed by its members, and the community has a central governing body with power over the entire territory. The members of a traditional community have the right to clear their customary forests to be controlled and used for the fulfilment of their individual needs and those of their families. Therefore, it is not possible for the rights held by customary law community members to be extinguished or frozen, provided that they meet the requirements of a traditional community as referred to in Article 18B(2) of the Constitution. (*Traditional Forest Community* case 2012: [3.13])

Constitutionally then, the State may not allocate land rights or concessions over forest areas – including for REDD+ purposes – if so doing would intrude on the rights of traditional communities who need access to those forests. Presumably, to award concessions over traditional land, the 'real' consent of those traditional communities must now be obtained.

However, a substantial legal stumbling block for traditional communities appears to remain: that is, in obtaining the formal 'recognition' required for them to be a 'traditional community' in the eyes of the law. Article 18B(2) of the Constitution requires that, to have the rights it provides, the community must 'still exist'. The Court appears to have interpreted this as requiring that the traditional community be formally recognised by the State. The Elucidation to Article 67 of the Forestry Law sets out the requirements for recognition as a traditional forest community: the community must exist in its traditional form; it must have institutions and a leader; occupy a defined area; have a legal infrastructure, including a customary law court to which its members adhere; and the surrounding forest area must be traditionally harvested to fulfil the daily needs of the community. Most significant is the requirement that a local government must, under Article 67(2), issue a regional regulation (*peraturan daerah*, or *perda*) to provide legal recognition of the community. The applicants in this case had met these requirements.[24] Other communities might not find this easy to achieve: some local governments are notorious for their lack of responsiveness to citizens' needs, making it difficult for traditional communities to convince their local governments to issue the regulation; or the recognition process might take significant time, allowing the government to issue a concession in the interim. Traditional communities are, however, likely to have the most difficulty convincing their local governments to formally recognise them by regulation if the local government itself wishes to award some type of permit, licence or concession over the very land that those communities use.

The Plantations Law *case (2010)*[25]

Although the *Traditional Forest Community* case has been described as a 'landmark' case (Pasandaran 2013), it is certainly not the first time the Court has invalidated statutory provisions for breaching Article 18B(2) of the Constitution. One earlier case was the *Plantations Law* case (2010). Article 9(1) of Law 18 of 2004 on Plantations (the 2004 Law) allowed the State to issue various rights over land to be used for plantations.[26] The applicants were individual citizens who lived and maintained land that had been designated for plantations under the 2004 Law and over which rights had been granted to commercial enterprises. This brought the applicants into regular conflict with plantation companies who held these rights and, the applicants claimed, often threatened them with Article 21 of the 2004 Law, which states: 'Every person is prohibited from performing acts that damage plantations or other assets, using plantation land without permission and or any other activity that causes impediments to plantation businesses.' The Elucidation to Article 21 defines 'using plantation land' as 'using plantation land or occupying it without the permission of the rightful owner in accordance with the law'. Article 47(1) of the 2004 Law provides for a maximum five-year prison sentence and Rp 5 billion fine for deliberate contravention of Article 21; Article 47(2) imposes those penalties by half for negligent contravention.

The applicants challenged the constitutionality of Articles 21 and 47, arguing that Article 21 was so loosely worded that it could be interpreted

---

24 The first applicants had had been recognised by Kampar Country Regulation 12 of 1999 on *ulayat* land; the second applicants by Lebak Regent Decision (430/Kep.318/Disporabudpar/2010).

25 Constitutional Court Decision 55/PUU-VIII/2010, reviewing Law 18 of 2004 on Plantations (*Plantation Law* case (2010)).

26 Article 9(2) requires that if there is *hak ulayat* land of an existing traditional community on the land needed for plantation, then before the rights referred to in Article 9(1) are granted, the prospective owner must first 'deliberate' with the *hak ulayat* holders to reach an agreement about the handover of land and compensation.

to encompass almost any activity. The Constitutional Court agreed, holding that this over-broadness caused legal uncertainty, which breached Article 28D(1) of the Constitution.[27]

The Court also found that Articles 21 and 47 violated Article 18B(2) of the Constitution. The Court noted that 'illegal' occupation of land (over which the State had granted concessions for plantations) had a long history in Indonesia, beginning during Dutch colonialism. The Dutch had granted concessions over large tracts of land that traditional communities had occupied for generations and had enacted laws that allowed land rights granted by the State to override traditional rights claimed by those communities. Many traditional communities were evicted without compensation and conflicts ensued. During the Japanese occupation in 1942–45, traditional communities were permitted to retake their plantation land, but were required to share proceeds with the Japanese government. After Indonesia declared its independence in 1945, the national government issued decrees that appeared to allow these traditional communities to continue occupying the former plantation land (although some of the new statutory rights issued by the State appeared to overlap with, and trump, some traditional land rights) (*Plantation Law* case 2010: 102–103). In this context, the Court decided that it was 'not appropriate' for Article 47(2) to impose criminal sanctions on a person who occupies land based on customary law, because customary law rights arise on the basis on occupation. According to the Court:

> It is appropriate that traditional community rights that have survived and accord with community developments within the framework of the unitary Republic of Indonesia be protected by statute. In this way, Article 18B will be capable of assisting traditional communities that are increasingly marginalised … To overcome the problem of ownership

27 Article 28D(1) states: 'Every person has the right to legal recognition, guarantees, protection and certainty which is just, and to equal treatment before the law.' The Court also held that Article 21 was so vague as to violate the *negara hukum*. This is commonly translated as the 'rule of law', but literally means 'law state' (Butt and Lindsey 2012). According to the Court (*Plantation Law* case 2010: [3.15.1]): '[a]cts that result in damage to the plantation' did not specify who would be caught by the provision. Would it capture a plantation owner who deliberately or negligently performed such acts themselves? And what would be encompassed by the word 'assets'? Need they be located on, or even associated with, the plantation? The Court also identified that 'other activities which cause disruption to plantation enterprises' is also extremely broad (*Plantation Law* case 2010: 104). The Court also posed various questions about this phrase. For example, could it be used to prosecute a person for late dispersal of a loan that had been agreed on by the plantation owner and the bank, leaving the plantation owner unable to purchase insect repellent, so that he could not prevent insects from causing damage to the plantation? Could a plantation owner be tried for neglecting his or her own plantation? According to the Court, Article 21 did not seem intended to capture those who performed acts such as these, let alone subject them to the criminal penalties in Article 47. Without much explanation and with no constitutional argument, the Court also opined that the acts covered in Article 21 were civil in nature, rather than criminal, and that disputes should be resolved through deliberation rather than by using criminal process.

disputes over plantation and *hak ulayat* land, the state should [act] in line with the Elucidation to Article 9(2) of the 2004 Law [in which it is stated that] traditional communities exist if the following five requirements are present: (a) a grouping in the form of a community, (b) institutions in the form of customary law leadership mechanisms, (c) a clear customary law area or region, (d) legal mechanisms, especially clear customary law dispute resolution mechanisms, (e) recognition by regional regulation. (*Plantation Law* case 2010: 103)

In this case, again the Court has, in effect, significantly diminished the constitutional rights it has upheld, for without this recognition the traditional communities cannot avail themselves of these rights. This recognition is likely to be difficult to obtain.

Coastal and Remote Areas Law *case (2010)*

Another predecessor to the *Traditional Forest Community* case was the *Coastal and Remote Areas Law* case (2010), discussed already in the context of State control and Article 33(3). We now return to analyse the case in the context of traditional communities. The Constitutional Court found that, by permitting concessions over coastal areas, the Law breached the traditional rights of communities on several constitutional grounds.

First, the Court seemed to imply within Article 33(3) an obligation on the State to protect pre-existing rights of traditional communities (*Coastal and Remote Areas Law* case 2010: [3.14.4]). According to the Court, when exercising control to achieve the greatest possible prosperity, the State was required to:

> [O]bserve existing rights, both individual and collective, held by customary law communities, the rights of customary law communities and other constitutional rights held by the community and which are guaranteed by the Constitution, such as the right to access to pass through, the right to a healthy environment, amongst others. (*Coastal and Remote Areas Law* case 2010: [3.15.4])

Bringing 'prosperity to the people' – the main purpose of the State exercising control under Article 33(3) of the Constitution – was, according to the Court, impossible if doing so would require the State to deprive people of natural resources, on which they rely for their subsistence needs and livelihoods.

Second, the Court held that, by permitting these concessions, the Law breached Article 33(4) of the Constitution and the State's obligation to advance public welfare and social justice under the Preamble. Article 34(2) states that 'the state is to develop a social security system for all people and is to empower weak and poor communities in accordance with human dignity.' According to the Court:

Providing these concessions breaches the principle of economic democracy [contained in Article 34(2)] which is based on the principle of togetherness and just efficiency ... [t]he principle of togetherness must be interpreted as meaning that when running the economy, including the management of natural resources for economic benefit, the broadest possible cross-section of the community must be involved and the prosperity of the people must improve. Management of natural resources must not merely observe principles of efficiency to obtain maximum profits, which can advantage a small group of capital owners, but must be able to increase the prosperity of the people in a just fashion. The exploitation of coastal areas and small islands by issuing concessions ... will result in [these regions] becoming concession areas controlled by large capital owners. By contrast, traditional fishing communities ... who rely for their lives and livelihoods on coastal natural resources will be excluded. In these conditions, the state has failed to fulfil its responsibility to run the national economy in a way that protects and provides justice to the people. More than this .. the issuance of these concessions will breach the principle of social justice for the entire Indonesian people as intended in the fourth line of the preamble to the Constitution. (*Coastal and Remote Areas Law* case 2010: [3.15.9])

Finally, the Court decided that the concessions would also reduce the level of community participation in determining how natural resources were used. Even though the Law requires that the community participate in the management of coastal areas, the Court held that this was insufficient to guarantee, protect and fulfil the rights of the community, with the more likely result being exclusion of the community (*Coastal and Remote Areas Law* case 2010: [3.15.8]).

Empowering local governments

Article 18(5) of the Constitution reads: 'Regional governments are to implement autonomy to the fullest extent, except in matters of government that are determined by statute to be matters for the Central Government.' In several cases, the Constitutional Court has invalidated statutory provisions that purport to grant the central government exclusive power to unilaterally designate zones in local government areas for particular purposes. In these cases, the Court has upheld challenges brought by local government officials complaining that the Constitution gives them rights over natural resource located in their regions and that, at the very least, they should participate in decision making about those resources. In this section, we discuss challenges to two statutes – the 1999 Forestry Law and the 2009 Mining Law.[28]

---

28 Law 4 of 2009 on Minerals and Coal Mining.

Forest Gazettal *case (2012)*[29]

This case was brought by several *bupati* (heads of district-level governments, commonly referred to as 'regents'). They challenged Article 1(3) of the Forestry Law, which appeared to allow the national government – in this case, the Forestry Ministry – to unilaterally designate (*menunjuk*) land as forest. As mentioned, the Forestry Law gives the State – primarily the Forestry Ministry – power to control activities in State forests, including by issuing and managing licences for logging, forest plantations and non-forestry uses (Wells et al. 2012: 5). Almost 90 per cent of forests had been designated in this way (Indrarto et al. 2012: 22–3). The *bupatis* argued, among other things, that Article 1(3) allowed the central government to allocate land as 'forest' even if that land was not, in fact, forested. The Forestry Ministry could also pursue those who illegally occupied or used land that it had designated as forest – a criminal offence under the Forestry Law – even if that land was not forest (Wells et al. 2012: 6). This left local people, including local governments themselves, beholden to the Forestry Ministry for permission to use the land so designated, including for community development and commercial endeavours. Article 1(3) had also frustrated local government attempts to devise and implement their own spatial plans, because the Ministry could undermine them by designating, as forest, land that the local government had allocated for another use.

The applicants argued that the Forestry Law contained a fundamental ambiguity. By contrast to Article 1(3), Articles 14 and 15 of the Forestry Law stipulated designation as merely one of four steps required for land to be 'confirmed' or 'gazetted' (*pengukuhan*). According to Article 15(1), the other three steps were adjustment, mapping and formal allocation.

In a short decision, the Constitutional Court found Article 1(3) to be partly unconstitutional. The Court decided that Article 1(3) would be invalid unless three words were excised from it, as follows: 'Forest areas are particular areas ~~designated and or~~ stipulated by the government to be maintained as permanent forest.' The Court's decision mentioned several constitutional grounds. First, the inconsistency between Articles 1(3) and 15 breached the constitutional guarantee of legal certainty provided in Article 28D(1).

Second, Article 1(3) breached the *negara hukum* concept, mentioned above, because it allowed the Forestry Ministry to unilaterally designate an area as forest. The Court explained that State administrative officials cannot act arbitrarily, but instead must act in accordance with the law and within their discretionary powers. According to the Court, the designation of an area as forest without involving forest-area stakeholders was authoritarian.

Third, the Court emphasised that Article 15(2) required spatial plans to be observed as part of the gazettal process, noting that this might uncover

---

29 Constitutional Court Decision 45/PUU-IX/2011, reviewing Law 41 of 1999 on Forestry (21 February 2012) (*Forest Gazettal* case (2012)).

pre-existing rights of individuals and traditional communities over the forest areas being designated. According to the Court, the Forestry Ministry could not designate a forest area where such pre-existing customary rights existed over the forest area to be designated (*Forest Gazettal* case 2012: [3.12.4]).

This decision appears to render invalid attempts by the central government to unilaterally designate areas as forest. It now appears that the central government is required to secure local government approval for the designation and to include the local government in the necessary demarcation and mapping. REDD+ projects require a licence from the Forestry Ministry 'for the purposes of sustainable forestry, ecosystem restoration, or afforestation', which can only be issued in forests classified as a permanent or production forest (Wells et al. 2012: 19). Local governments might, pointing to this decision, now contest areas in their territories being allocated as such forests and may insist on their being used for non-forestry uses – such as lucrative palm oil plantations. In terms of REDD+, the net effect of this decision is to increase difficulties in obtaining licences over forests for REDD+ purposes. Presumably, the decision also hands more bargaining power to local governments in negotiations with the central government, and perhaps even donors, about financial and other returns from REDD+ 'to compensate for the forgone short-term development it replaces' (*Forest Gazettal* case 2012: [3.12.4]).[30]

Mineral and Coal Mining Law *case (No. 1) (2010)*

This case was brought by various individuals as well as environmental and human rights NGOs. They challenged provisions of the 2009 Mining Law,[31] which allowed 'mining areas' (*wilayah pertambangan*) to be formally determined by the national government – after it had coordinated with regional

30 The Court issued a decision with similar effects in the *Mineral and Coal Mining* case (No. 2) (2012) (Constitutional Court Decision 10/PUU-X/2012 reviewing Law 4 of 2009 on Mining). Similarly to Constitutional Court Decision 032/PUU-VII/2010, reviewing Law 4 of 2009 on Mining (*Mineral and Coal Mining Law* case (No. 1) (2010), this was a constitutional challenge to the Mining Law provisions, which granted the national government power to stipulate mining areas (*wilayah pertambangan*) after 'coordinating' with local governments, 'consulting' with the national parliament, and observing (*memperhatikan*) the opinion of the community. Unlike *Mineral and Coal Mining* case (No. 1), which was brought by traditional community members who claimed that the provisions failed to meaningfully include the community in decisions to stipulate areas for mining, this case was brought by the Kutai Timur *bupati*. He argued that the provisions breached his rights as head of the district to exercise broad autonomy under Articles 18(5) and 18A(2) of the Constitution and to obtain justice in the exploitation of natural resources for the prosperity of the Kutai Timur community under Articles 33(2) and (3). The Court agreed, essentially amending the impugned provisions so that a mining area needed to be stipulated (*ditentukan*) by both the national and regional government in which the mining area was located. In other words, it appears no longer sufficient for the national government to discuss proposed mining areas with local governments, but to ignore any objections. Rather, it appears that the national government must secure the agreement of relevant local governments to areas within the jurisdictions being designated for mining.

31 Articles 6(1)(e), 9(2), 10(b), 14(1), 14(2), 17, 162, 136(2) and 171(1).

governments, consulted with the national parliament and observed (*memper-hatikan*) the opinion of the community. The applicants complained that local and traditional communities, occupying or otherwise having traditional rights or entitlements over the land, were not involved in the process. This, they claimed, threatened their constitutional rights and entitlements over land and other natural resources, the right to a good and healthy environment and the right to not be forcibly removed (*Mineral and Coal Mining Law* case (No. 1) 2010: [3.13]). The applicants also pointed out that, by continuing to occupy and use resources in a government-stipulated 'mining area', communities faced criminal sanctions for preventing or disrupting mining business activities (*Mineral and Coal Mining Law* case (No. 1) 2010: [3.9]). The applicants asked the Court to require the central government to obtain the signatures of all affected people in the proposed 'mining area', along with approval from the local government and national parliament, before designating that area (*Mineral and Coal Mining Law* case (No. 1) 2010: [3.13]).

After referring to several of its previous decisions,[32] the Court observed that the Indonesian people had, through the 1945 Constitution, authorised the State to exercise 'control' over natural resources to achieve the greatest possible prosperity of the people; but also required the State, when exercising that control, to actively respect, protect and fulfil the economic and social rights of citizens. According to the Court, however, the provisions requiring coordination with regional governments, consultations with the national parliament and observation of the opinions of the community were 'mere procedural formalities' that obscured these State obligations (*Plantation Law* case 2010: [3.13.1]). The Mining Law did not, for example, specify whose opinion from among the community needed to be sought. The Court decided that the central government could not act unilaterally and arbitrarily when determining mining areas and that observing (*memperhatikan*) the opinion of the community 'must be interpreted as imposing an obligation, rather than a discretion, upon the government'. In other words, the State was required to consider the opinions of the community likely to be affected by the proposed mining area. According to the Court, these community opinions constituted a 'form of control' or a 'control function' over the government to ensure that it respected the constitutional rights of citizens. These included the right to live prosperously, to have a place to live and a good and healthy environment, to have personal property and not have property arbitrarily expropriated (Articles 28H(1) and (4)) (*Plantation Law* case 2010: 139–40). The Court rejected the applicants' suggestion that the written consent of all affected persons must be obtained before an area could be allocated for mining, pointing out that forgery would likely be a problem. Instead, the Court held that the community needed to be 'directly involved', but left the mechanisms for community involvement, including whose opinion would be sought, to be regulated by the government.

---

32 Including the *Electricity Law* case (2003) and the *Investment Law* case (2007).

Conclusion

Indonesia's Constitutional Court can be expected to influence in a significant way the manner in which a national REDD+ scheme is conceived and implemented in Indonesia. Presuming that the scheme is statutory, it is highly likely that traditional communities, NGOs such as AMAN, or even local government authorities will bring constitutional challenges against the scheme before the Constitutional Court, as they have in cases discussed in this chapter. It is highly likely that the Court will uphold these applications if they employ similar grounds to the cases already decided. It is possible, and perhaps likely, that the Court will extend its reasoning in cases involving particular types of natural resource to cases involving other resource types. For example, it is likely that the Court's decision in the *Oil and Natural Gas* case would be applied to forests if the same arguments were to be made in subsequent cases about forestry.

On the one hand, these cases appear likely to make large-scale REDD+ projects far more legally complicated and expensive to introduce. They make it difficult, if not impossible, for the national government to nominate large tracts of forests for REDD+ purposes, at least without complex negotiations with local governments and traditional communities. They are also likely to limit the duration of any concession over land on which REDD+ projects might run and also preclude the government from signing away absolute control of forests, particularly to private enterprises. In this context, we note that under current Forestry Ministry regulations, REDD+ projects may run for up to 30 years and are extendable.[33] If the national legislature were to enact statutory provisions providing for a similar duration, allowing upfront extensions, or both, then it is likely that the Court would invalidate those provisions if challenges to them are lodged. If the issuance of a concession leads citizens, including customary law communities, to lose access to the resources they need for their livelihoods, then the Court is likely to invalidate the statutory grounds on which the concessions can be awarded. On the other hand, these cases should help Indonesia comply with some of the safeguards discussed in earlier chapters and its international legal obligations to its indigenous peoples.

References

Afghani, M. M. A. (2013) 'The elements of 'state control', *Jakarta Post*, 14 January.

Butt S. (2014) 'Traditional land rights before the Indonesian Constitutional Court', *Law, Environment and Development Journal*, 9(2).

Butt, S. and T. Lindsey (2008) 'Economic reform when the Constitution matters: Indonesia's Constitutional Court and Article 33', *Bulletin of Indonesian Economic Studies*, 44(2): 239.

---

33 Article 13 of Minister of Forestry Regulation 30/Menhut- II/2009 on Mechanisms for Reducing Emissions from Deforestation and Forest Degradation.

Butt, S. and T. Lindsey (2012) *The Indonesian Constitution: A contextual analysis*, Oxford: Hart Publishing.

Butt, S. and F. Siregar (2013) 'The BP Migas case: Implications for the management of natural resources', *Journal of Energy & Natural Resources Law*, 31(2): 107.

Indrarto, G. B., P. Murharjanti, J. Khatarina, I. Pulungan, F. Ivalerina, Justitia et al. (2012) *The context of REDD+ in Indonesia: Drivers, agents and institutions*, Bogor, Indonesia: CIFOR and ICEL.

Pasandaran, C. (2013) 'Constitutional Court annuls government ownership of Customary Forests', *Jakarta Globe*, 17 May.

Wells, P., N. Franklin, P. Gunarso, G. Paoli, T. Mafira, D. R. Kusumo et al. (2012) 'Indonesian Constitutional Court Ruling Number 45/PUU-IX/2011 in relation to Forest Lands', *Daemeter/TBI Indonesia/Makarim & Taira S.*

10 Conclusion

The protection of tropical forests has been a subject of global concern for over three decades. However, tropical deforestation is part of a much longer history of land use change, with over 800 million ha of grasslands, savannah, woodlands and forests converted to cropland and pastures between 1860 and 1980 (Christoff and Eckersley 2013: 132). In total around one-third of the global land surface has been transformed from its natural state for agricultural purposes (Ellis 2011: 1025). This process has had a major impact on many landscapes across the planet. We now know that it has also had significant and ongoing effects on the global climate. Between 1750 and 2011, CO_2 emissions from deforestation and land use change contributed around 180 gigatonnes of CO_2 to the atmosphere, around one-third of the carbon emitted to the atmosphere from human activities (IPCC 2013: 12). Undoubtedly, curbing deforestation and other land use change that releases CO_2 will be vital to preventing dangerous global warming in the twenty-first century and beyond.

As was the case with the fractious debates over the Amazon and other tropical forests in the 1990s, there was a risk that the inclusion of forestry governance in global climate policy would once again pit the north against the south. However, contemporary discussion of global forestry governance has taken a more positive and less divisive turn for several reasons. First, there is growing understanding that unmitigated climate change will spare no State and that developing countries such as Indonesia are highly vulnerable to rising temperatures and sea levels. They have as great an interest in combatting climate change as any nation in the north. Second, whereas global debates around forestry in the 1990s were largely concerned with biodiversity preservation and involved no substantial compensation from north to the south to achieve this objective, REDD+ explicitly seeks to create financial incentives for developing countries by linking land use emissions abatement to the international carbon market. REDD+ has therefore been a 'game changer' (Murdiyarso et al. 2012: 680) and has rapidly emerged as the dominant global legal framework through which States are seeking to control greenhouse gas emissions from land use change in developing States.

The fundamental challenge for global policymakers is that for all the good intentions behind REDD+ it will not be effective unless and until it can be transformed into domestic law and policy on the ground and can command the support and participation of local communities (Murdiyarso et al. 2012). There is an extensive body of work on the concept of REDD+ and on the opportunities for implementing REDD+ in several countries, including Indonesia. This book has built on these accounts of REDD+ by examining in depth the legal and institutional environment for natural resource management in Indonesia. In this final chapter, we identify the main conclusions to emerge from this research.

The opening chapters of the book (Chapters 1, 2 and 3) demonstrated that, from fairly modest beginnings, REDD+ has undergone very significant development as an international policy instrument for mitigating climate change. One of the most important developments has been the increasing acknowledgement of the interests and needs of forest communities. Indonesia has been a very willing participant in the evolution of international climate change law and policy, especially in relation to REDD+, which holds significant promise for linking Indonesia with global carbon markets and delivering financial returns to the Indonesian economy. However, there are further opportunities for Indonesia to become engaged in international climate law and this can by catalysed by clearer identification of Indonesia's vulnerability to global warming and the range of mitigation options open to the government, most notably REDD+.

Chapter 4 of the book provided an overview and analysis of the institutional and legal environment for REDD+ in Indonesia. Indonesia's post-Soeharto democratisation and decentralisation processes have improved public participation in government and the rule of law. However, the devolution of authority to Indonesia's provinces, districts and cities has posed new challenges for natural resource policymaking at a national level. Indonesia is not unique in this respect; similar challenges have been confronted by federal countries in both the south (e.g. India) and north (e.g. the United States). However, in Indonesia, the transfer of authority from the centre to subnational governments has not been accompanied by effective mechanisms to resolve overlaps and disputes between different levels of government. Indonesia's decentralisation therefore remains a work in progress, with the Supreme Court and Constitutional Court key forums in which jurisdictional disputes over REDD+ will be played out over coming years. Chapter 4 also showed that endemic corruption remains a significant hurdle. Decentralisation has presented newly empowered subnational officials with unprecedented opportunities to gain private benefits from public decision making, including decisions to awarding licenses to exploit natural resources such as forests. Fragmentation in governance and ongoing corruption, especially in relation to forestry issues, significantly complicates the implementation of REDD+. It also means that Indonesia cannot make credible commitments in international climate negotiations to provide a domestic environment for REDD+ programmes on which investors can safely rely.

Chapter 5 discussed the implementation of international norms in Indonesian law, a topic that has received surprisingly little analysis among commentators despite being critical to the internalisation of REDD+ and other aspects of international climate law in the Indonesian legal system. As one of the world's most populous countries with a fast growing economy, Indonesia is an important participant in the international legal system and is party to major regional and global treaties, including the 1982 United Nations Convention on the Law of the Sea and the 1992 United Nations Framework Convention on Climate Change and its 1997 Kyoto Protocol. Formally, Indonesia appears to adopt a 'monist' approach to international law, regarding treaties and customary law as automatically entering domestic law. However, in practice, there is much uncertainty and domestic implementation is normally required for giving effect to treaties in Indonesian law. We argued that given the complexity of REDD+, and the uncertainties surrounding international law in Indonesian law, REDD+ must be given a firm legal basis in national legislation if it is to have any real prospects for success.

Chapter 6 examined the national regulatory framework for REDD+ in Indonesia, beginning with an overview and assessment of the Forestry Law. The key Presidential regulations of relevance to REDD+ were also analysed, including the 2013 Moratorium Continuation Instruction which extended a moratorium on the issuing of new forest concessions over natural forest or peatlands within certain conservation and production forests. However, the most important REDD+ regulations have been those issued by the Forestry Ministry, including the 2009 REDD+ Mechanisms Regulation, the first attempt to provide an overarching legal framework for REDD+ in Indonesia, and the 2012 Forest Carbon Management Regulation, the most recent regulation addressing REDD+ demonstration activities (which have had a poor record of performance to date). On paper at least, this suite of legislation and regulation, along with a Presidential institution (the REDD+ Taskforce), has the capacity to advance the implementation of REDD+ in Indonesia. However, as Chapter 6 further indicated, the legal framework is indeed only that – and requires further development and elaboration to provide the level of certainty required to encourage large-scale REDD+ projects in Indonesia.

An additional challenge is the clear demarcation of jurisdictional authority, an issue examined in detail in Chapter 7. Decentralisation and democratisation in Indonesia have resulted in the dispersion of power across the three arms of government and at national and subnational levels and this has profound implications for the implementation of the REDD+ laws discussed in Chapter 6. Laws from regional and national institutions have proliferated, with many of them inconsistent or in outright conflict with each other, but the Indonesian legal system provides no clear demarcations of jurisdictional authority. However, without clarity as to jurisdictional hierarchy it is not possible to determine whether REDD+ regulations can, in fact, take effect on the ground. We examined two evolving mechanisms that can be used to provide greater certainty as to institutional authority for natural resource management, namely, bureaucratic review by the central government (of disputes between the central and regional governments) and judicial review in the Supreme Court. These mechanisms are flawed in several respects and more problematically do not extend to inter-agency competition and disputation at the national level. A concrete example of this is the conflict between Forestry Ministry Decree 36 on Permit Procedures for Carbon Sequestration and Law 17 of 2003 on State Finance administered by the Finance Ministry. The dispute goes to the heart of the successful implementation of REDD+ as it concerns which of the Forestry or Finance Ministry should have authority to determine revenue-sharing arrangements from REDD+ activities. Given that REDD+ is fundamentally a financial instrument for forest conservation and climate mitigation, this continued uncertainty over profit sharing serves as a major impediment to REDD+.

Against this background of jurisdictional unpredictability, Chapter 8 examined the potential for litigation relating to REDD+ activities in the Indonesian court system. Although litigation may, in a sense, be undesirable as it will be an obvious symptom of underlying disputes over forest protection and REDD+ regulation, judicial decisions may provide a helpful means for the incremental development of the law and may also bring clarity on those issues that will require a more robust legislative basis at a national level. Despite systemic weaknesses in the Indonesian legal system (including corruption), there are positive signs of movement towards the rule of law, which is important not only for resolving jurisdictional disputes but also to provide redress for individuals and entities that may be adversely affected by unlawful activity by governments and other actors in relation to REDD+ activities. In this respect, the 2008 Freedom of Information Law may assist in improving public scrutiny of natural resource management in Indonesia and lead to better public administration of REDD+ if civil society groups fully utilise Information Commission processes. Nevertheless the Freedom of Information Law has is limits and is subject to significant exemptions (for instance can 'natural resources information' be restricted from public access?). As such it is unlikely, in its current form, to provide the degree of transparency and accountability that is truly needed to evaluate fully the success or otherwise of REDD+ projects.

The Indonesian Constitutional Court, established in 2003, is likely to be a key forum for the resolution of disputes over REDD+. Chapter 9 was devoted to examining the development of the Court as an institution and the main areas of its jurisprudence of relevance to REDD+. Several dimensions of the Court's jurisprudence have major implications for REDD+. One is that that the Court has sought to protect the rights of traditional communities in disputes concerning natural resources, which carries the implication that any REDD+ legislation will need to take seriously the safeguards provisions developed at the international level in the climate regime and give them meaning within the context of Indonesian constitutional law. Moreover, the

Court has sought to protect the rights and interests of provincial and local governments, which means that unilateral national legislation on REDD+ that does not evince a serious attempt to engage subnational governments make be struck down as unconstitutional.

This book has shown that Indonesia is a crucible for one of the most vital areas of natural resource management that has implications not only for sustainable development in the world's fourth most populous country, but has global ramifications for climate stability. It is difficult to envisage a scenario in which dangerous global warming can be avoided unless the carbon sinks in Indonesian forests and peatlands are largely preserved intact. The international community therefore has a major stake in the success of REDD+ in Indonesia, a reality underscored by Norway's significant support of forest conservation initiatives through the concession moratorium. However, for REDD+ to be successful, there needs to be far greater understanding and acknowledgment of the legal and institutional hurdles to its implementation in Indonesia. This work has sought to make a contribution to developing this understanding, by providing an in-depth assessment of the Indonesian legal and political system as it has relevance to REDD+.

References

Christoff, P. and R. Eckersley (2013) *Globalization and the environment*, Lanham, MD: Rowman & Littlefield.

Ellis, E. (2011) 'Anthropogenic transformation of the terrestrial biosphere', *Philosophical Transactions of the Royal Society A*, 369: 1010–35.

Intergovernmental Panel on Climate Change (IPCC) (2013) 'Summary for policymakers', in T. F. Stocker, D. Qin, G.-K. Plattner, M. Tignor, S. K. Allen, J. Boschung et al. (eds) *Climate change 2013: The physical science basis. Contribution of Working Group I to the Fifth Assessment Report of the Intergovernmental Panel on Climate Change*, Cambridge: Cambridge University Press.

Murdiyarso, D., M. Brockhaus, W. D. Sunderlin and L. Verchot (2012) 'Some lessons learned from the first generation of REDD+ activities', *Current Opinion in Environmental Sustainability*, 4: 679–85.

Index